STAIRS

THE BEST OF
Fine Homebuilding

STAIRS

THE BEST OF
Fine Homebuilding

The Taunton Press

Cover Photo: Charles Miller

Taunton
BOOKS & VIDEOS

for fellow enthusiasts

First printing: 1996
Second printing: 1998
Printed in the United States of America

A FINE HOMEBUILDING Book

FINE HOMEBUILDING® is a trademark of The Taunton Press, Inc.,
registered in the U.S. Patent and Trademark Office.

The Taunton Press, Inc.
63 South Main Street
P.O. Box 5506
Newtown, Connecticut 06470-5506
e-mail: tp@taunton.com

Library of Congress Cataloging-in-Publication Data

Stairs : the best of Fine homebuilding.
        p.   cm.
     Includes index.
     ISBN 1-56158-131-3
     1. Stair building — Miscellanea.   2. Stairs — Miscellanea.
   3. Staircases — Miscellanea.   I. Taunton Press.   II. Fine
   homebuilding.
   TT5667.S73   1996
   694'.6 — dc20   95-47080
                                                                   CIP

CONTENTS

INTRODUCTION

STAIRBUILDING is the epitome of the carpenter's art. Often the centerpiece of a house, a fancy staircase combines the complex geometry of roof framing with the rigorous demands of furniture-grade joinery. But even the most mundane set of cellar stairs must be carefully laid out and assembled. If one step in a staircase is even ½ in. higher or lower than the others, you can feel it, likely because it trips you up. And consequently such a staircase does not comply with building codes. There's good reason why rookie carpenters are relegated to running baseboard in closets, not sent to build the stairs.

But stairbuilding shouldn't be approached with trepidation. You simply need to have a healthy respect for the demands of the work. And you need good information.

This book brings together 30 articles from the back issues of *Fine Homebuilding* magazine. Written by builders, architects and engineers about their own work, these articles will guide you through projects ranging from straight and simple stairs to curved and complex ones.

A word of caution, though. The building codes that govern stairs are changing all the time, so not all of the projects featured here comply with current national codes. If you're building any staircase, be sure to check with your local building department. And remember, as Tracy Kidder wrote in his book *House*, "Stair-making carpenters are like school crossing guards or trainers of seeing-eye dogs. They take on one of society's small sacred trusts."

—*Kevin Ireton, editor*

Designing and Building Stairs

Stairways can be minimal or very elaborate, but they're all based on simple geometry and accurate finish work

by Bob Syvanen

Today, few people can afford the time and expense of building classical 19th-century stairways. Tastes have changed too, and as a result, stairs have become simpler. But the principles of stair design and construction are the same as they've always been, and so are the skills that the builder must bring to the task. If modern stairs aren't ornate, they still remain a focal point of a house—your work is out there for all to see.

One of the problems builders face in cutting stairs is getting rusty. Unlike the master stairbuilders of the past, on-site carpenters typically build only a few stairways a year. Fortunately, designing a stair and laying out the stringers uses the same language and concepts as roof framing. Fitting treads and risers, on the other hand, is nothing more than simple, if demanding, finish work.

Basic stair types—The simplest stairway is a cleated stairway, which relies on wood or metal cleats fastened to the carriages to support the treads. You could use a cleated stairway for a back porch or cellar, but count on repairing the ever-loosening cleats. Wood cleats are typically 1x4s, and screwing them in is a big improvement over nails. But angle-iron cleats will last longer.

Another open-tread stairway—one without risers—uses dadoed carriages. The treads have ½ in. or more bearing on the inside face of the carriages, and they are either nailed or screwed in place. I use a circular saw, and set the depth of cut to half the thickness of the carriage, to make the parallel cuts. Then I clean up the bottom of the dado with a chisel. A router and a simple fixture built to the stair pitch and clamped to the carriage will also do a quick, neat job.

In the past, dadoed carriages were used mostly for utility stairs for porches, decks and the like, but more and more I'm asked to build open-tread oak stairs that have to be nearly furniture quality. One type, the stepladder stair, can be used when limited space doesn't allow any other solution (for more, see p. 14).

On finished stairs, the dado is usually stopped (that is, its length is limited to the

Stairbuilding in its highest forms requires the conceptual skills of a roof framer and the fitting talents of a cabinetmaker. But simple open tread runs like the basement stair at left only require understanding the basics.

width of the tread), and squared up at the end with a chisel. If you begin the dado on the front of the carriage, leave the tread nosing protruding slightly. If you start the dado at the back of the carriage, the treads will be a little inset. They can look nice either way.

Other stairways use cut-out carriages. The simplest of these is a typical basement stair where the rough carriages are exposed, and risers are optional. The most complicated is a housed-stringer stairway. It is based primarily on patient and accurate work with a router. The stringer is mortised out along the outline of the treads and risers with a graduated allowance behind the riser and tread locations for driving in wedges. Adjustable commercial fixtures or wooden shop-made templates are used to guide the router. Although the first stairway that I built by myself on my own had a housed stringer, I won't try to give complete instructions here.

A finished stairway that still requires patient finish work, but is much less tedious to build, uses cut-out carriages hidden below the treads, and stringboards or skirtboards as the finish against the stairwall. The treads and risers are scribed to the skirts. The example I'll be using to describe final assembly is one of these that also has an open side, which uses a mitered stringer. This side of the stair requires a balustrade—handrail, balusters, and newel posts—but that's a separate topic.

Designing a stairway—No matter what style stairway you want to build, the design factors that you'll need to consider are comfort, code, safety and cost. Comfort gives the ideal conditions for good walking, and code dictates what you can and cannot do. Safety is largely a matter of common sense, and cost limits your grand ideas.

Although comfort is very much a subjective notion, there are three objective factors to consider in designing a stair—stair width, headroom and the relationship between the height of the riser and the width of the tread. Of the three, stair width is the easiest to deal with. Most building codes require utility stairs to be at least 2 ft. 6 in. wide and house stairs to be 3 ft. from wall finish to wall finish, but 3 ft. 6 in. feels a lot less restrictive. However, if the stairway gets beyond 44 in. in width, most building codes require a handrail on each stair wall.

Headroom is not quite as simple, although most codes agree that basement stairs need a minimum of 6 ft. 6 in., and house stairs need at least 6 ft. 8 in. This measurement is made from the nosing line (for a definition, see *headroom* in the glossary on the next page) to the lowest point on the ceiling or beam above. A lack of headroom is most noticeable when you're going down the stairs, because you are walking erect and bouncing off the balls of your feet. Ideal headroom allows you to swing an arm overhead going downstairs, but this requires a clearance of nearly 7 ft. 4 in. Headroom can be increased by enlarging the size of the stairwell, decreasing the riser height, or increasing the width of the treads (tread

width is the distance from the front to the rear of the tread).

Certain combinations of riser and tread are more comfortable than others. Most codes set limits—a maximum rise of 8¼ in. and a minimum tread width of 9 in.—but these are based on safety, not comfort. A 7-in. rise is just about ideal, but it has to be coupled with the right tread width.

The timeworn formula for getting the tread width and riser height in the right relationship is: riser + tread = 17½ (a 7-in. riser + a 10½-in. tread = 17½ in.) Another rule of thumb is: riser × tread = 75 (7 × 10½ = 73.5, which is close enough). Still another formula is: two risers + one tread = 24 in. I've always found the first formula the easiest. All of them establish an incline between 33° and 37°. This creates a stairway that is comfortable for most people.

More considerations—Each tread should project over the riser below it. This projection, or nosing, should be no more than 1¼ in. and no less than 1 in. In open-tread stairways, a tread shouldn't overlap the tread beneath it by more than ½ in. The nosing adds to the area where your foot falls, but doesn't affect the rise-run dimensions. The top of a handrail should be between 30 in. and 34 in. above this nosing, with a 1½-in. clearance between the handrail and the wall. If the top of the stairway has a door, use a landing at least as long as the door is wide.

Keep in mind that people aren't the only things moved up and down stairs. I once lived in an old Cape Cod house that had a stairway with 10½-in. risers, 6½-in. treads, and not too much headroom. You could negotiate it if you exercised a little caution, but moving heavy furniture up and down was another story.

The size of a stairwell is based on the riser and tread dimensions, and on how much headroom you need. A typical basement stair can be gotten into a rough opening 9 ft. 6 in. long by 32 in. wide. A main stairwell should be a minimum of 10 ft. by 3 ft.

Framing a stairwell isn't complicated and is usually defined in the local building code. If the long dimension is parallel to the joists, the trimmer joists on each side are doubled, as are the headers. Similar framing is required if the long dimension is perpendicular to the joists (drawings, above right). If the header isn't carried by a partition below, it will have to be designed for the load.

Making the calculations—Careless measuring can get you in a lot of trouble when you're building stairs. Although the initial figuring may seem a little theoretical, you'll soon be doing some fussy finish work based on these calculations and the resulting carriages. First, check both the stair opening and the floor below for level. If either is out of level, determine how much. You'll have to compensate for it later. This problem occurs most often with basement slabs.

There are many ways to lay out stairs; the following system teamed with a pocket calcu-

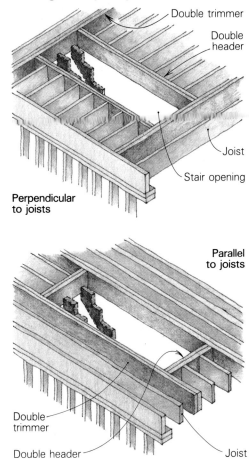

Framing stair openings

Double trimmer

Double header

Joist

Stair opening

Perpendicular to joists

Parallel to joists

Double trimmer

Double header

Joist

lator works well for me, even though I have to scratch my head to recall what I did last time. Start with the measurement that has already been determined, the finished floor-to-floor height. I'll use 108 in. in this case. Then, just for a starting point to get you close, divide by 14, the average number of risers used in residential stairs. This gives a riser of 7.71 in., which is a little high. Adding another riser will reduce this measurement some: 108 in. ÷ 15 = 7.20 in. Sounds good. You now know how many risers you'll be using and how high they are.

To get the width of the treads (remember, this means the front-to-back measurement of each step), use the rise-plus-tread formula in reverse: 17.5 in. − 7.2 in. = 10.3 in.

All stairs have one more riser than treads. This is because the floors above and below act as initial treads, but aren't a part of the stair carriage calculations. You'll have to keep reminding yourself of this when you lay out the carriages. I don't know a good way of remembering this, and I have resorted to drawing a sketch of a couple of steps and using it to count the difference. In the example we're using, then, there are 14 treads and 15 risers.

The last calculation is the total run—the length of the stairway from the face of the first riser to the face of the last riser. This is simply the total number of treads multiplied by their width: 14 × 10.3 in., or 144.2 in. With this figure you can check to see whether the stairway will fit in the space available. Although I now use a calculator for the math, my main tool used to be a stair table like the

A Glossary of Stair Terms

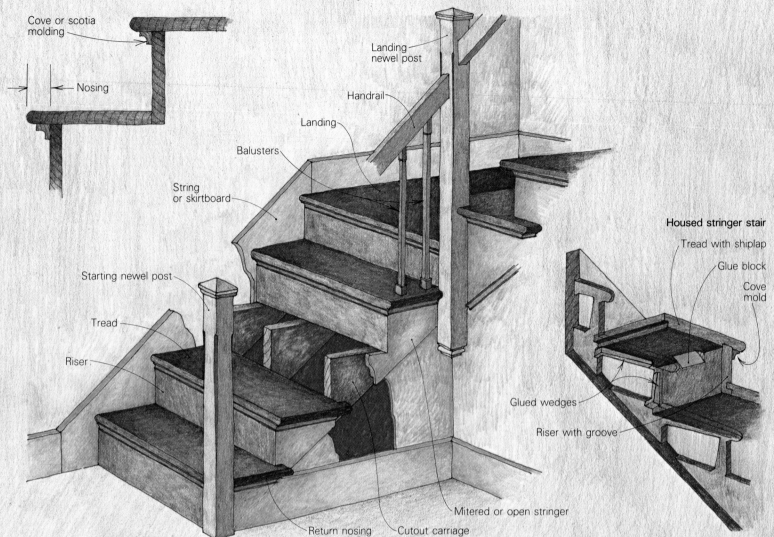

Cove or scotia molding

Nosing

Landing newel post

Handrail

Landing

Balusters

String or skirtboard

Starting newel post

Tread

Riser

Housed stringer stair

Tread with shiplap

Glue block

Cove mold

Glued wedges

Riser with groove

Mitered or open stringer

Return nosing

Cutout carriage

Balusters—The posts or other vertical members that hold up the handrail, usually two per tread.

Balustrade—The complete railing, including newel posts, balusters and a handrail. Most of these parts are available as stock finished items at lumberyards.

Carriages—Also called *stair stringers, stair horses* or *stair jacks.* They are the diagonal members that support the treads. Carriages can either be *finish stringers* or *rough stringers*—for an outside stairway, or for an inside stairway hidden from view. *Rough carriages,* whether they are *cut-out carriages* or just *dadoed* or *cleated stairs,* are made of 2x10 or 2x12 softwood lumber. Finish stringers are usually made of ¾-in. or 1⅛-in. stock. They either can be cut out (an *open or mitered stringer*) or routed (a *housed stringer*).

Closed stairway—Stairs with walls on both sides. In this case a *wall stringer,* whether it is a *housed stringer* or just a *stringboard,* is nailed to each wall. Closed stairways use handrails, not a balustrade.

Finished stairway—Any of several interior stair types that have risers, treads, stringers and a handrail or balustrade.

Handrail—This rail runs parallel to the pitch of the stairs. It's held by balusters or brackets.

Headroom—The vertical distance from the lowest point of the ceiling or soffit directly above the stair to the *nosing line,* an imaginary diagonal connecting the top outside corners of treads. Most codes require at least 6 ft. 8 in. for stairs in living areas, and 6 ft. 6 in. for basement utility stairs.

Housed stringer—The profile of the treads, nosing and risers is routed into a finish stringer. Extra room is left for wedges to be driven and glued in between the stringer and the treads and risers. Rabbeted and grooved risers and treads are also used.

Landing—A platform separating two sets of stairs.

Newel post—The large post at the end of the handrail. There is a *starting newel* at the base of the stairs, and a *landing newel* at turns.

Nosing—The rounded front of the tread that projects beyond the face of the riser 1 in. to 1¼ in. In the case of *open-tread stairways,* it shouldn't exceed ½ in. In most cases, the nosing is milled on the tread stock. On open stairways, a half-round molding called *return nosing* is nailed to the end of the tread.

Open or mitered stringer—This is a cut-out finish stringer used in open stairways. The treads carry over the stringer, but the vertical cut-outs on the carriage are mitered with the risers at 45°.

Open stairway—This can be open on one or both sides, requiring a balustrade. In finished stairways, the open sides will use a *mitered* or *open stringer.*

Rise—The height of each step from the surface of one tread to the next. Just as in roof framing, this measurement is sometimes called the *unit rise.* Many codes call for a maximum rise of 8¼ in. The height of the entire stair, from finished floor to finished floor, is the *total rise.*

Riser—Describes the rise of one step. It is also a stair part—the vertical board of each step that is fastened to the carriages. Risers for a *housed stringer stair* are rabbeted at the top to fit the tread above, and grooved near the bottom for the tread below. Other stairs use 1x square-edged stock. *Open-tread stairs* don't have risers.

Run—Also called *unit run,* this is the horizontal distance traveled by a single tread. A 9-in. run is the code minimum for main stairs. *Total run* is

the measured distance from the beginning of the first tread to the end of the last tread—the horizontal length of the entire stairway.

Stairwell—The framed opening in the floor that incorporates the stairs. Its long dimension affects how much headroom the stair has.

Stringboard—Diagonal trim, not used to support the treads, that is nailed to the stair walls. Finished treads and risers butt these. Often called *skirtboards, backing stringers,* or *plain stringers.*

Tread—It is both the horizontal distance from the face of one riser to the next, and the board nailed to the carriages that takes the weight of your foot. Exterior stairs typically use 2x softwood treads. Interior stairs use either 1⅛-in. hardwood stock milled with a rabbet and groove to join it to the risers, or 1³⁄₁₆-in. square-edged stock. Both are usually nosed.

Winder—Wedge-shaped treads used in place of a landing when space is cramped, and a turn is required in the stairway. Many building codes state that treads should be at least the full width of the non-winder treads, 12 in. in from their narrow end; or that the narrow end be no less than 6 in. wide.

—*Paul Spring*

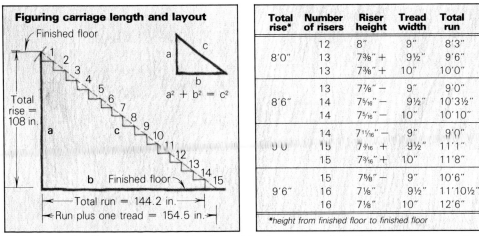

Figuring carriage length and layout

Finished floor

Total rise = 108 in.

$a^2 + b^2 = c^2$

Total run = 144.2 in.

Run plus one tread = 154.5 in.

Total rise*	Number of risers	Riser height	Tread width	Total run
8'0"	12	8"	9"	8'3"
	13	7⅜"+	9½"	9'6"
	13	7⅜"+	10"	10'0"
8'6"	13	7⅞"−	9"	9'0"
	14	7 5/16"−	9½"	10'3½"
	14	7 5/16"−	10"	10'10"
9'0"	14	7 11/16"−	9"	9'0"
	15	7 3/16"+	9½"	11'1"
	15	7 3/16"+	10"	11'8"
9'6"	15	7⅝"−	9"	10'6"
	16	7⅛"	9½"	11'10½"
	16	7⅛"	10"	12'6"

*height from finished floor to finished floor

Stair geometry. The stair described in the text is shown above left. Plugging total run and total rise into the Pythagorean theorem gives the required carriage length. Based on an ideal riser and tread, the stair chart, above right, gives the number of risers and total run for a given total rise.

one above. I still use one for quick reference in the planning stage. If the run is too long, make the treads narrower, or eliminate a riser and a tread. Either way, you'll need to run a new set of calculations.

Layout—To lay out the carriages, you first need to know what length stock to buy. If you've got a calculator, it's easiest to use the Pythagorean theorem $(a^2 + b^2 = c^2)$. But you must add the width of an extra tread to your total run to get enough length for the bottom riser cut (drawing, above). In this case, a is the total rise of 108 in., and b is the total run of 144.2 in. plus a 10.3-in. tread. The hypotenuse, c, is 188.5 in. So you'll have to buy 16-footers to allow for cutting off end checks, and avoiding large knots with the layout.

Most cut-out carriages are 2x12s because you need at least 3½ in. of wood remaining below the cut-outs for strength; 4 in. is even better. Douglas fir is the best lumber for the job because of its strength. You will need a third, or center, carriage if the stair is wider than 3 ft. with 1½-in. thick treads, or wider than 2 ft. 6 in. with 1⅛-in. thick treads.

Once you have marked the edge of one of the carriages with the 188.5-in. measurement, you are ready to lay out. I step off equal spaces with dividers (photo top right) before marking the riser and tread lines with a square. Some carpenters simply step off the cut-out lines with a square, but I don't like the accumulated error you can get this way. A deviation of more than ¼ in. between the height of risers or the width of treads can be felt when walking a flight of stairs. For this reason, it's also a violation of code.

The dividers I use are extra long. You can improvise a pair by joining two sticks with a finish nail for a pivot, and a C-clamp to hold them tightly once they're set. The easiest way to find the spacing is to locate 7.2 in. (the riser height) and 10.3 in. (the tread width) on a framing square, and set the divider points to span the hypotenuse, which is 12.56 in. (a strong 12½ in.). No matter how careful you are in setting the dividers, it will take a few trial-and-error runs before you come out to 15 even spaces. Once you do, mark the points on

the edge of the carriage. These represent the top outside corner of each tread, less the profile of the nosing.

To draw the cut-out lines, use either a framing square or a pitchboard. Most carpenters use the square, but a pitchboard can't get out of adjustment. You can make one by cutting a right triangle from a plywood scrap. One side should be cut to the height of the riser, and the adjacent side to the tread width. A 1x4 guideboard should be nailed to the hypotenuse. Align it with the marks on the carriage and use it to scribe against.

If you are using a square, set it on the carriage so that the 90° intersection of the tongue and body point to the middle of the board. Along the top edge of the carriage, one leg should read the riser increment and the other, the tread increment. Use either stair-gauge fixtures (stair buttons) or a 2x4 and C-clamps to maintain the correct settings when the square is in position against the edge of the carriage. Then, holding the square precisely on the divider marks, scribe the cutlines for each 90° tread-and-riser combination (photo second from top).

Dropping the carriages—One of the most difficult things about stairs is adjusting the carriages for the different thickness of floor finish, which can throw off the height of the bottom and top risers. Any difference should be subtracted from the layout after the tread and riser lines are marked, and carefully double-checked before you do any cutting. What you marked on the carriage is the top of the treads, but since you will be nailing treads to the carriages, you need to lower the entire member enough to make up for the difference. This is called *dropping the carriages*. If they sit on a finished floor, such as a concrete basement slab, the bottom riser will need to be cut shorter by the thickness of a tread (drawing A, next page, top right). This will lower the carriage so that when the treads and upper floor finish are added, each step will be the same height. The bottom riser will have to be ripped to a narrower width. If the treads and floor finishes are of equal thickness (B), and the carriage sits on the subfloor at the bottom,

Laying out and cutting the carriages. Starting with the top photo, Syvanen uses large dividers to mark the intersections of tread and riser lines on the front edge of the carriage. From these marks, he scribes the cut-out lines using a framing square fitted with stair-gauge fixtures. With a circular saw, he begins cutting out the carriages by notching the bottom riser for the kickplate that will anchor it to the floor. Syvanen uses a handsaw to finish the cutting. Cut-out carriages are usually made of 2x10s or 2x12s, since there should be at least 3½ in. of stock between the bottom edge of the carriage and the cutout.

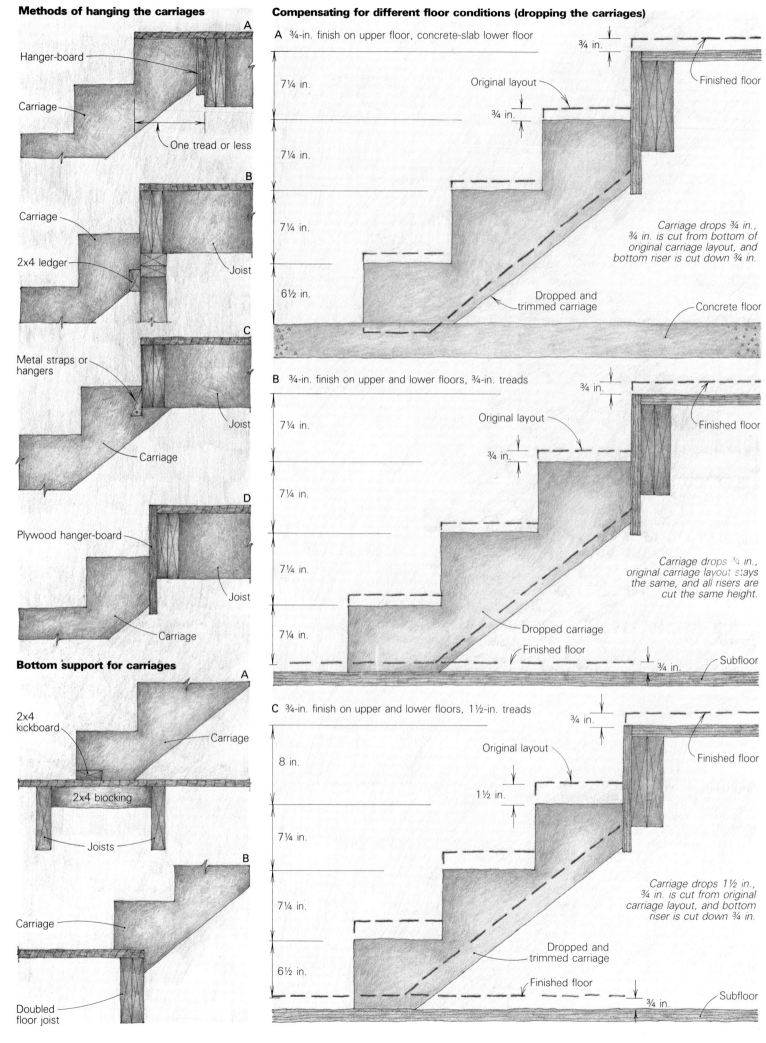

Methods of hanging the carriages

A
Hanger-board
Carriage
One tread or less

B
Carriage
2x4 ledger
Joist

C
Metal straps or hangers
Joist
Carriage

D
Plywood hanger-board
Joist
Carriage

Bottom support for carriages

A
2x4 kickboard
Carriage
2x4 blocking
Joists

B
Carriage
Doubled floor joist

Compensating for different floor conditions (dropping the carriages)

A ¾-in. finish on upper floor, concrete-slab lower floor

¾ in.
Original layout
Finished floor
7¼ in.
¾ in.
7¼ in.
7¼ in.
6½ in.

Carriage drops ¾ in., ¾ in. is cut from bottom of original carriage layout, and bottom riser is cut down ¾ in.

Dropped and trimmed carriage
Concrete floor

B ¾-in. finish on upper and lower floors, ¾-in. treads

¾ in.
Original layout
Finished floor
7¼ in.
¾ in.
7¼ in.
7¼ in.

Carriage drops ¾ in., original carriage layout stays the same, and all risers are cut the same height.

Dropped carriage
Finished floor
7¼ in.
¾ in.
Subfloor

C ¾-in. finish on upper and lower floors, 1½-in. treads

¾ in.
Original layout
Finished floor
8 in.
1½ in.
7¼ in.

Carriage drops 1½ in., ¾ in. is cut from original carriage layout, and bottom riser is cut down ¾ in.

7¼ in.
Dropped and trimmed carriage
6½ in.
Finished floor
¾ in.
Subfloor

Nearly complete, the finish stair at left is missing only its balusters, handrail, and molding under the nosing. The newel post is mortised into the first tread for stability. This stair uses both an open stringer and treads and risers that butt-join the skirtboards or stringboards that are nailed to the wall. Above, kraft paper protects the completed oak treads from construction traffic. The open stringer is mitered to the riser, and the treads overhang the stringer by the depth of the return nosing with its scotia molding beneath. The cut-out carriage that actually supports the treads is hidden behind the finish stringer and drywall blocking.

no change will have to be made for the risers to be equal.

A more confusing condition is when the treads are thicker than the finished floor (C). At this point, I usually draw a four-riser layout, at any scale, on graph paper to figure how much of a drop I need to make, and if the bottom riser needs to be narrower.

How the cut-out carriages are attached once they are raised in the stairwell also may require adjustment at the top and bottom of the carriages (drawings, opposite page, bottom left. At the bottom, I like to use a 2x4 kickboard nailed to the floor at the front edge of the riser (A). If there is a stair opening below, the carriages can be cut to fit around the upper corner of the framing (B). Stairs take a beating, and should be well secured.

At the top, the header joist usually acts as the uppermost riser, but sometimes, the floor will extend a full or partial tread width from the framing (drawing A, opposite page, top left). A 1x4 or 2x4 ledger board can be nailed to the framing (B), and if so, the carriages must be notched to fit it. Metal angles or straps can be used if the carriages aren't exposed (C). I like a hanger board because it is quick, neat and strong. I nail the carriages to a line on a piece of plywood, a riser's distance from the top. I then raise the whole business as a unit and nail it in place (D).

Give the carriage a trial fit before sawing it out. I make only the horizontal cut that rests on the floor and the vertical cut that leans against the framing at the top before trying

the carriage in the stairwell. If you are really unsure, use a 1x10 trial board. With the carriage in place, you can easily check your layout. The treads should be level from front to back, and the carriage should fit on both sides of the opening. Also make sure that the risers will all be the same height once the treads and finish floor are installed.

If everything checks out, what's left is just cutting and fitting. With this basic layout you can produce the cut-out carriages that are needed for the stairway shown above, you can dado the carriages for let-in treads, or you can just nail cleats to the layout lines. For cut-out carriages, use a circular saw as far as you can, and finish them off with a handsaw held vertically so as not to overcut the line and weaken the cut. You can nail the triangular cutouts to a 2x6 for a third stringer if the budget is tight. Use the completed carriage as a pattern to trace onto the other 2x12, and then cut the pencil line to get an exact duplicate.

Treads and risers—On a closed stairway, the cut-out carriages sit inside the finish wall stringers, which are called skirtboards, or strings. These are usually 1x10, and should be nailed hard against the wall so that the snug fit of previously installed risers and treads isn't spoiled by the skirtboards spreading when newly scribed boards are tapped into place. They should be installed parallel to the nosing line, and as high as possible without exposing any wall where the riser and tread meet. Don't nail the cut-out carriages to the

skirtboards on a closed stairway, or the mitered stringer to the outside carriage on an open stair. Instead, hold the carriages about 3 in. away from the walls, so that they are bearing only at their tops and bottoms. This keeps the treads and risers from splitting as a result of nailing too close to their ends. Skirtboards and risers can be made of pine to ease the budget, but the best treads are oak. The standard thicknesses for treads are 1⅛ in. and 1³/₁₆ in.

I like to rip all of my treads and risers to width before beginning the assembly. Keep the risers a hair narrower than what's called for. Crosscut both risers and treads to 1 in. longer than the inside dimension between the skirt boards. This allows them to fit at a low enough angle to get a good scribe and still have a little extra to cut off. If the stair is open on one side, the treads will have to be rough-cut long enough to leave a ½-in. scribing allowance on the closed side, and some overhang on the open side, which gets a return nosing. The risers will need at least a 45° miter to mate with the open stringer. Use a radial arm saw or handsaw for this.

Stair assembly usually begins at the bottom. The first two risers are fit and nailed, and then the first tread is pushed tightly against the bottom edge of the riser for scribing. For the stair pictured above, I first had to cut the open, or mitered, stringer. It was pine, and was laid out like the carriages with the exception of the vertical cuts, which extend beyond by the thickness of the riser material to form

Stepladder stairway

A cleated or dadoed stairway at an angle of from 50° to 75° is considered a stepladder stairway, or ships's stairway. The rise on these ladders can be from 9 in. to 12 in. As the angle of the carriages increases, the rise increases. A 50° angle should have about a 9-in. rise. A 75° angle should have about a 12-in. rise.

As with all other stairs, the relationship between the height of the riser and the width of the tread is important. But in this case, their relationship is reversed. The tread width on stepladder stairs will always be less than the riser height.

If you know what riser height you'll be using, the easy formula for calculating the tread width is this: tread width = 20 − 4/3 riser height. If you use a 12-in. rise, then the tread will be 4 in. You can also rearrange this formula to solve for riser height (riser height = 15 − 3/4 tread width) if you know what tread width you want. This is useful if you are limited in how far out into a room the base of the stair can come.

A simple way to lay out a stepladder stairway is to lean a 2x6 or 2x8 carriage at about 75° in the stair opening. Lay a 2x4 on edge on the floor, and scribe a pencil line across its top edge to transfer a level line from the floor onto the carriage. Mark the vertical cut, at the top, in a similar manner.

Make these cuts and set the carriage back in place. Measure the floor-to-floor height and divide by 12 in. to get the number of risers. You'll end up with a whole number and a fraction. For instance, if your total rise is 106 in., the number of risers will be 8.83. That's close enough to call it 9 risers. Divide this back into the total rise of 106 in. to get an accurate riser dimension of 11.77 in., or about 11¾ in.

The easiest way to lay out the carriages is to make a story pole using dividers set at about 11¾ in. to step off nine equal segments within the 106 in. It might take a few tries adjusting the dividers to get it to come out just right. You can even find the riser dimension without the math by making a few divider runs up and down the story pole.

With the carriage in place, mark each tread from the story pole. A level line at each mark locates the treads. A bevel square set at the correct angle will also work. The treads can be cleated or let in with a dado. —B.S.

an outside miter. I'm most comfortable making these cuts with a handsaw. Keeping this angle slightly steeper than 45° allows the faces to meet in an unbroken line, without any interference at the back of the joint. The miters were predrilled, glued and nailed with 8d finish nails (photo previous page, right).

The treads are initially cut to overhang the open stringer, by the same dimension as the nosing. Then a cross-grain section is cut out so that a mitered corner is left at the outer edge. This accommodates a return nosing. I also like to put a piece of nosing at the rear of the tread where it overhangs, although you can just round off the back end of the return nosing. If the mitered stringer doesn't snug up perfectly under each tread, don't worry. The crack will be covered by the cove or scotia molding that runs under the nosing. It's more important to get good bearing for the tread on the cut-out carriage, and a little shimming or block-planing will help here.

Once beyond the open side of the stair, you will be fitting treads and risers on a closed stairway, scribing to the skirtboard on each side. With a 1-in. allowance for scribing, set the scribers at ½ in. for the first side. Set the tread or riser with the side you are going to scribe down in place on the carriage and against the skirtboard. The other end will ride high on the other skirt. If you are working on a tread, make sure it is snug to the riser along its entire length. Risers should sit firmly on the carriages to get an accurate scribe. I use a handsaw to back-bevel the cut on risers, but I keep the cut square at the front of the treads where the nosing protrudes, and then angle it the rest of the way. If necessary, use a block plane to make sure the cut fits.

Next, get the inside dimension at the back of the tread or lower edge of the riser, depending on which you are fitting. A wood ruler with a brass slide works well here. Transfer this dimension to the board you're working on, set the scriber to the remaining stock, and mark the board. Cut the boards the same way you did the first time. Cut carefully, remembering that you can always plane it off, but you can't stretch it. However, don't make the cut too strong either. Trust your measure.

Careful cutting and fitting are important with this kind of stair, and a little glue, and some nails and wedges in the right place work wonders as time goes on. Risers and treads are nailed to the carriages with 8d finish nails through predrilled holes. A sharpened deheaded 8d finish nail chucked tightly into an electric drill makes a snug hole every time.

Stairs that aren't made from rabbeted stock (like housed-stringer stairs) can be kept together if you drive three or four 6d common nails through the bottom of the riser into the back of the tread. This should be done as you fit your way up the stair. Two 1x1 blocks, 2 in. long, glued behind each step at the intersection of the upper edge of the riser and the front of the tread will cut down a lot on movement too. Any gaps between the carriage and treads should be shimmed from behind with wood wedges to eliminate squeaks. □

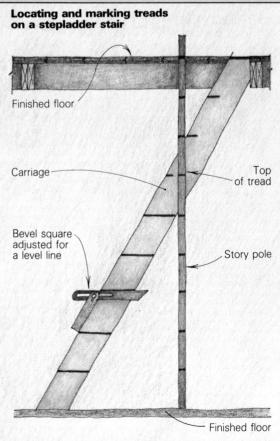

Locating and marking treads on a stepladder stair

Finished floor

Carriage

Top of tread

Bevel square adjusted for a level line

Story pole

Finished floor

Design Guidelines for Safe Stairs

Getting the details right might save a life

by Gregory Harrison

What could be simpler than building a stair? Consider, for a moment, statistics from the National Safety Council and the Consumer Product Safety Commission: over 13,000 Americans die each year because of falls, with 3,800 killed on stairs each year. About 2.5 million stair falls are tallied each year, and 800,000 of them result in injuries that require professional medical care. However, because injuries can be treated in other than emergency rooms, it's probable that this figure is low—some published estimates put the true figure at more than 2 million serious injuries. Simply put, stairs are the most dangerous architectural feature in a house. It doesn't take much thought to realize that this problem is one that architects, engineers and builders create, and that it's also one they can readily solve.

In the course of my work as a safety engineer, expert witness and design consultant, I have studied thousands of stairs and hundreds of fall accidents. If you design, specify, sell or build stairs, I suspect that the subject of this article won't immediately arouse your interest. But I submit that a lawsuit for wrongful death or severe bodily injury caused by "your" stairway would.

The role of the codes—The philosophy of model building codes has been to relax stair safety requirements for one- and two-family dwellings, even though that's where the majority of fatalities occur. Why this is so is not clear, but the code agencies in general have a record that reflects a resistance to code changes affecting personal safety. Perhaps this is because professional safety engineers aren't typically consulted, or because the process of code development reduces the output to the lowest common denominator. The bottom line, though, is this: codes contain technical errors, and no code can anticipate every condition in the field, so don't rely on a building code or standard alone to guide the design or construction of stairs. Compliance with the codes is not a defense against negligence per se, contrary to what most architects and builders think. As one well-known trial attorney put it, "Codified negligence is still negligence."

Basic stair design—Based on my experience, I believe a safe stairway, whether in commercial or residential use, should have the following characteristics (top drawing, following page).

1. Reachable, continuously graspable, and structurally stable handrails on both sides, with intermediate handrails as required;
2. Properly proportioned risers and treads with close tolerances;
3. Slip-resistant treads and nosings;
4. Adequate lighting, appropriately located and controlled;
5. Guardrails (and toeboards on steps if open on the side);
6. General compliance with the NFPA Life Safety Code; and
7. At least three steps.

There are other factors that, while not pertaining directly to the design of stairways, can figure into the safety equation. The stairs should be properly maintained, for example, with no loose treads or wobbly handrails. And there shouldn't be any environmentally triggered factors that could distract the stair user, such as mirrors or HVAC ducts that might suddenly blow air at a stair user.

Most of the requirements above seem to be the stuff of common sense, but several are so important that I'd like to explain them in detail: handrail design, tread/riser design, slip-resistance, short runs, and guardrails.

A safe handrail is crucial—A graspable handrail is *the* most important characteristic of a safe stair because, regardless of how a fall is caused, a good handrail offers the stair user a last chance to reduce the impact of a fall. A handrail also serves as a visual signal that a change of elevation exists, and it provides a continuing support for use by the elderly and by children. To be functional (remember form follows function?), a handrail must be *graspable*. One well-known architect and stair-safety researcher, Jake Pauls, has suggested what he calls, "The Acid Test of Graspability." Because the purpose of a handrail is to provide a secure grip for people who may be taking urgent and desperate action to prevent death or a disabling accident, Pauls suggests that designers be required to hang from two sections of their proposed handrails, one grasped with each hand, and maintain that grasp while suspended over a vat of acid. If his proposal were invoked, there would certainly be fewer handrails constructed of 2x4s, 2x6s, 2x8s and large-diameter pipes (which are particularly ungraspable by human beings).

To be graspable means that you can curl

your fingers and thumb around and underneath the handrail. After investigating hundreds of stair-fall accidents, I've come to the conclusion that the proper design of handrails has evidently been beyond the grasp of many designers and builders, especially in homes.

The National Fire Protection Association publishes Standard No. 101, called the *Life Safety Code* (LSC), and it contains, in Chapter 5, extensive architectural and engineering criteria relative to the safe design of stairs and ramps. I recommend that you get a copy (National Fire Protection Association, Batterymarch Park, Quincy, Mass. 02269; 800-344-3555). The 1988 LSC handily summarizes handrail graspability (see the sidebar on p. 17). The lack of any handrail, or the existence of an oversized one, represents a very unsafe and dangerous condition. The dimensions of a proper handrail aren't difficult to understand (drawing, p. 17).

It would seem obvious that the handrail should extend along the full length of a stair, but this isn't always done (bottom drawing, following page). A stair design that allows steps to extend beyond the reach of the handrail is not a wise idea, particularly because those first steps are where most of the accidents happen.

I think builders and architects, out of professional duty and plain common sense, ought to provide handrails for *all* stairways, including those with only one or two steps (notwithstanding the lack of such a requirement in the current one- and two-family building codes). What could be cheaper than installing an attractive wood or brass handrail that stands ready to reduce or eliminate injuries for many years to come? Based on my professional accident investigations in residences, I'm convinced that much suffering and misery could easily be eliminated with the addition of proper handrails. Think of a proper handrail as analogous to a seatbelt or an airbag in a car.

Riser/tread design—During the design of any stair, risers and treads should be carefully detailed to result in the proper geometry and size. Human-factors research at the National Bureau of Standards (now known as the National Institute of Standards and Technology) has shown that foot travel clears a riser by as little as $3/8$ in. Most codes require $3/16$-in. maximum variation between adjacent risers and treads. That means that the steps should be carefully and solidly constructed: a stable step

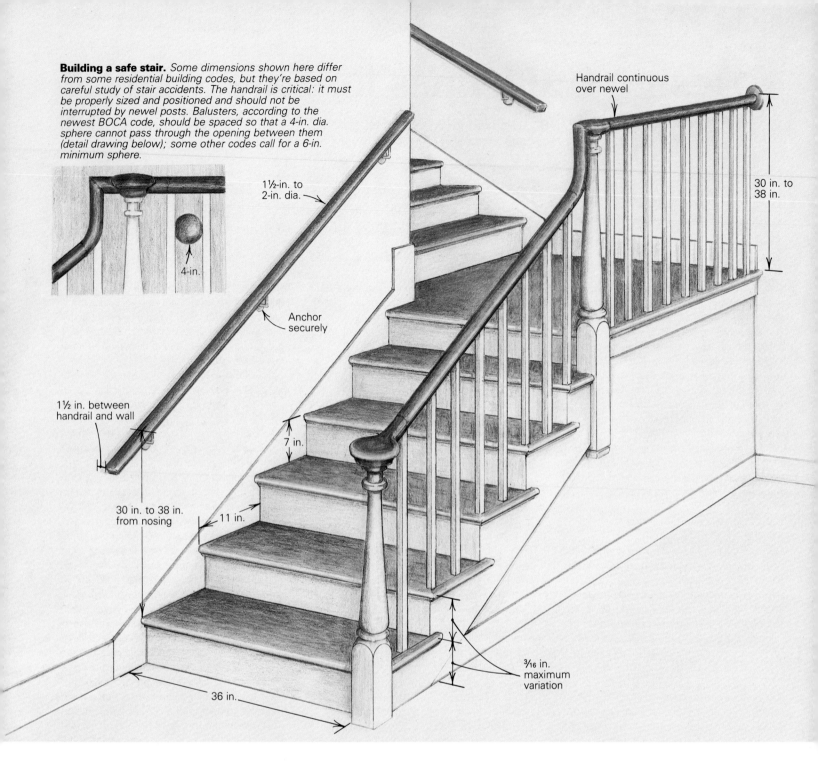

Building a safe stair. *Some dimensions shown here differ from some residential building codes, but they're based on careful study of stair accidents. The handrail is critical: it must be properly sized and positioned and should not be interrupted by newel posts. Balusters, according to the newest BOCA code, should be spaced so that a 4-in. dia. sphere cannot pass through the opening between them (detail drawing below); some other codes call for a 6-in. minimum sphere.*

4-in.

Handrail continuous over newel

1½-in. to 2-in. dia.

30 in. to 38 in.

Anchor securely

7 in.

1½ in. between handrail and wall

30 in. to 38 in. from nosing

11 in.

36 in.

³⁄₁₆ in. maximum variation

---

is less likely to warp or twist out of position and affect these tolerances. Finally, codes usually require that within the overall stairway, risers and treads not vary more than ³⁄₈ in.

The ideal step dimensions, based on my review of the safety literature, is 7 in. for the riser and 11 in. for the tread, hence the "7-11 step." The minimum riser height by code is 4 in. and the maximum is about 8 in. Although I know of no code maximum for treads, unusually deep treads cause an abnormal gait that can cause missteps and a fall. The CABO One and Two Family Dwelling Code allows treads to be only 9 in. deep, which is ridiculous because almost all adult feet with shoes exceed this measurement.

One formula for expressing step geometry was put forth in 1672 by Francois Blondel of the Royal Academy of Architecture in Paris and takes the form of:

$$2 \times R + T = 24 \text{ in. to } 25 \text{ in.}$$

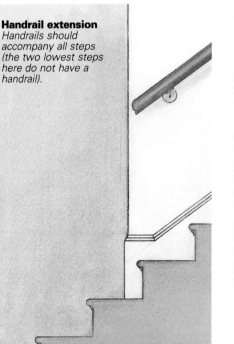

Handrail extension
Handrails should accompany all steps (the two lowest steps here do not have a handrail).

Other formulas found in older editions of various building codes include:

$$R + T = 17 \text{ to } 17.5 \text{ in.}$$
$$R \times T = 70 \text{ to } 75 \text{ in.}$$

These formulas have been eliminated from modern codes because they are far from perfect and exclude some safe designs; you should not rely on them.

Slip-resistant surfaces—The slip-resistance of stair treads is an important consideration in providing safe stairs, especially when it comes to the nosing area. Careful selection of materials is the best way to ensure safety, and certain materials are inherently dangerous as a stair surface: smooth concrete, polished marble, terrazzo, ceramic tile (especially glazed tile), painted wood, and most any other highly polished hard surface, including oak with a polyurethane finish.

During a recent show of upscale new model

Drawings: Bob Goodfellow

homes, I noticed that most of the builders had installed highly polished marble or ceramic tile foyers and steps. Attendants were posted everywhere to give the visiting public verbal warnings about the very slippery conditions. With the steep price of each home in mind, I inquired as to whether or not the same attendants came with the house so that my guests would receive the same warning and, hence, reduce my liability. No deal. Not only would these steps require frequent cleaning cycles and constant scratch control, but numerous serious falls would surely occur over the years. Glazed tiles and polished stone belong on the walls, not on floors and especially not on steps. Given these conditions, however, there's all the more need for very graspable handrails on both sides, regardless of stair width.

Short and dangerous—Short flights of stairs are often employed to accommodate changes in elevation of less than 21 inches, both indoors and outdoors (such as for decks and patios). But short flights are inherently dangerous. In fact, flights of one or two risers are so undesirable that the Life Safety Code committees and model codes have, in the past, attempted to prohibit their use. They wanted to substitute ramps for stairs with fewer than three risers, or where changes of elevation were less than 21 inches.

The main reason short flights are dangerous is that many people do not see the steps until they have already begun to fall. Also, most stairway-fall accidents occur on the first or second step ascending or descending; if the flight only has two steps, a user will always be in the zone of relatively high hazard in either direction of travel. The LSC recommends the avoidance of short flights, but states that if steps are used in a short flight, the tread should be a minimum of 13 in. and that each step location should be readily visible. You would be well advised to read section A-5-1.6.2 of the 1988 NFPA LSC Standard No. 101 or the *Encyclopedia of Architecture* (Vol 4, 1989) concerning this subject.

The most elegant way to prevent falls on short flights of stairs is to replace them with a ramp. If your client insists on having short flights, or if you've inherited the problem, there are ways to reduce the incidence of falls.

Because the primary problem is visual, you should include all possible visual cues to the presence of steps. Here are some possibilities:

—Slope handrails at the same pitch as the stairs.

—Use contrasting surfaces (a polished brass rail against a dark wall, for example).

—Install textured or patterned wall coverings so they follow the angle of the steps, thereby highlighting the change.

—Provide relatively intense lighting to highlight the steps.

—Remove geometric-print carpet from stairs (it can camouflage the presence of steps).

—Provide step nosings with illuminated strips embedded in their top surface; one manufacturer of such a product is Diamond Metal Ltd. (80 Colonnade Rd,, Nepean, Ont K2E 7L2; 613 226-1123).

—If the room has a normal ceiling height, slope it downward over the steps, parallel to the stair angle. The change of ceiling plane will draw attention to the change in level.

—Provide tactile cues. The use of hardwood treads, for example, on all steps (including the top landing) can alert a pedestrian approaching from a carpeted area of an impending change underfoot. The sensation of stepping from padded carpet to a hard surface tends to cause one to look down to see what the change is about.

Guardrails—Balusters, railings, Plexiglas, tempered glass and other such guardrails serve a useful function: they keep you from falling *off* stairs and landings. Most codes require guardrails to be 42 in. to 44 in. high in commercial settings. The 36-in. height usually prescribed for residential applications is, I believe, inadequate. Based on a study performed by engineer Elliott Stephenson and backed by the American Academy of Pediatrics, the BOCA code responded efficiently (to their credit) and recently reduced the baluster spacing from 6 in. to 4 in. (see detail drawing, facing page). A 6-in. baluster spacing will allow 950 out of 1,000 children less than 10 years old to pass through, so it's not a very effective barrier. In stark contrast, few children, except those less than a year old, can pass completely through a 4-in. wide opening. The American Institute of Architects and the National Association of Home Builders have been less than enthusiastic about this and certain other safety-related changes, according to Mr. Stephenson. It has also been my professional experience, based on more than 20 years of meeting with architects, that they usually do not appreciate the possibility of falls. I'm sure many architects believe that I do not appreciate the aesthetic aspects of my suggestions, either.

Spirals, winders and landings—Other types of stairs such as spiral stairs and winders are not as safe as straight-run stairs because of the varying tread width. Thus, they require a *very* good handrail. Consult the NFPA LSC for details concerning risers and treads.

Straight-run stairs are not as safe as stairs that have landings—if you fall, you'll travel a bumpy road to the bottom of the stair. Landings shorten a fall, and they should be at least as long as the stair is wide. If a stair rises more than 12 feet (a vaulted interior space, for example) the code requires a landing.   □

Gregory Harrison, P. E., has more than 20 years' experience as a safety, fire protection and civil engineer.

Handrail graspability

"Handrails should be designed so that they can be grasped firmly with a comfortable grip and so that the hand can slide along the rail without encountering obstructions. The profile of the rail should comfortably match the hand grips. For example, a round profile such as is provided by the simplest round tubing or pipe having an outside diameter of 1½ in. to 2 in. (3.8 cm to 5 cm) provides good graspability for adults. Factors such as the use of a handrail by small children and the wall fixing details should be taken into account in assessing handrail graspability. The most functional as well as the most preferred handrail shape and size is circular with a 1.5 in. (3.8 cm) outside diameter (according to

research with adults). Handrails used predominately by children should be designed at the lower end of the permitted dimensional range. It should be noted that handrails are one of the most important components of a stair; therefore, design excesses such as oversized wood handrail sections should be avoided unless there is a readily perceived and easily grasped handhold provided. At all times in handrail design it is useful to remember the effectiveness of a simple round profile that permits some locking action by fingers as curl around the handrail."

—From the Life Safety Code, National Fire Protection Association, 800 Batterymarch Park, Quincy, Mass. 02269.

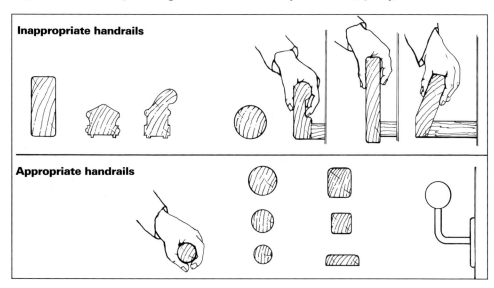

Inappropriate handrails

Appropriate handrails

A Veteran Stairbuilder's Tools and Tips

Modified tools and custom jigs expedite staircase assembly

by Michael von Deckbar-Frabbiele

If you've ever poked around in an old toolbox, you're likely to have pulled out some strange-looking gizmo that, even after careful scrutiny, confounded all by its presence. Such tools typically elicit the comment "I wonder what they used *that* for?" As a woodworker who specialized in stairbuilding, I made plenty of special tools and jigs that'll probably end up as curiosity-provoking what's-its. These tools and jigs made my job easier, and, as someone once told me, it's not doing hard work that makes one a master, it's making hard work easier. So let's take a look at a stairbuilder's gizmos and at some techniques that should help you do a few things in your work: reduce effort, increase productivity, elevate your degree of accuracy and, ultimately, increase your profit.

Reground spade bits work better

Judicious grinding of spade bits makes them bore holes in diameters between stock sizes. Beveled corners prevent tearout.

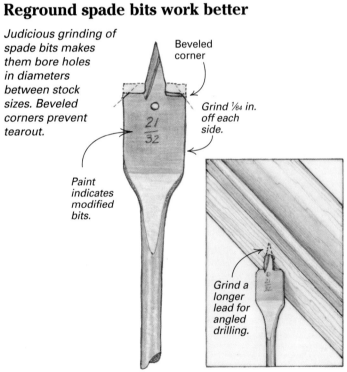

Beveled corner

Grind ¹⁄₆₄ in. off each side.

²¹⁄₃₂

Paint indicates modified bits.

Grind a longer lead for angled drilling.

The worst thing in the entire world that can befall a stair man is to hear his just-installed treads squeak. The second-worst thing is to have the spindles rattle. Children love it; they never seem to tire of running up and down the balustrade with an arm extended, slapping the spindles to produce a staccato to which their youthful nervous systems seem immune.

For spindles not to rattle, they have to fit perfectly in their holes. To make a perfect fit, I modify common spade bits. It's easy to grind down the bits by ¹⁄₃₂ in. (or ¹⁄₆₄ in. on each side). So, for example, instead of jumping from ⅝ in. to ¹¹⁄₁₆ in., you'll have a bit that's ²¹⁄₃₂ in.

Another hint: Because the newly modified bits will have their former sizes stamped on them, it's important to paint new numbers on the sides of each bit. Once, a carpenter who was setting a balustrade went into my toolbox without my knowledge and bored 35 holes with what he thought was a ¾-in. bit.

Another thing about spade bits: When boring at an angle, say, into an oak handrail, a spade bit's 90° corners tend to tear out chunks of wood as they start a hole. Grinding off the bit's corners makes a clean cut by producing a scraping action as the bit spins into wood. I grind a long lead on some bits to make them useful for grinding holes in steep handrails. The long lead establishes the bit in its hole before the shoulder engages wood.

Shortened level fits on tread

Leveling across the width of a tread is easier with a 10-in. level cut from a larger level. Extension caps screwed to each end of the level accommodate cupped treads.

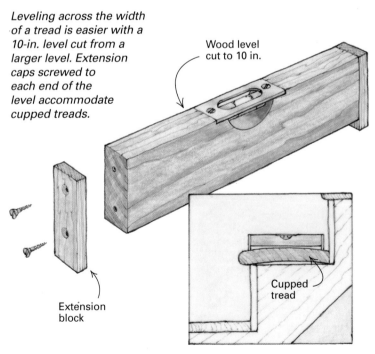

Wood level cut to 10 in.

Extension block

Cupped tread

When leveling treads from front to back, it's handy to have a small level. You could buy a small bullet level, but I've never found one that didn't seem like a toy. I cut down a larger wood level to 10 in. and screwed small extension blocks of wood on the ends. The blocks extend past the bottom of the level because sometimes a tread isn't exactly flat, so the extension blocks allow the body of the level to clear the belly of a chubby tread and

Drawings: Bob Goodfellow

give an accurate read. For years I used a longer 2-ft. level when I was setting treads, but I was constantly bumping into the level where it was protruding past the tread and knocking it down the staircase. Aside from the fact that the tumbling level dented and dinged up the treads during its descent, I got pretty tired of buying new levels because the old ones got knocked out of whack.

Gauge measures shim thickness

A tapered shim gauge, cut from a scrap of wood and calibrated by the 16th of an inch, makes a handy gauge for determining shim thickness.

Fill in with colored marker for easy reference.

In many cases, the stair carriage (or horse, or rough stringer) is built by the framing carpenter, who is long gone by the time you come along to make a silk purse out of an old sow's ear. No offense to the framers; many do a great job. Nonetheless, it is in your best interest to build the rough stringers yourself, or at least check them with level and rule before you bid on finishing the stairs. It once took me three days with a reciprocating saw, a firmer chisel and a mass of shim stock to straighten out a circular-stair string (pun intended) before I could trim it out.

I make a shim gauge out of a scrap of wood. I measure along both edges of the shim and make marks every $\frac{1}{16}$ in. in thickness. I take a marker and color in every other segment. After I've got the gauge made, I rip shims of different thicknesses and keep them on hand. Some people use shingles as shims, but their tapered profile gives them uneven bearing; the surface to be shimmed only hits the high point of the shingle.

To use the gauge, I simply slip it under the tread or behind the riser (inset drawing above) that needs shimming and tap it in until the tread is level or the riser is plumb. I note the mark on the gauge, remove the gauge and replace it with one of my precut shims. Once you have the shim in place, the difficult work is done, and then it's just a matter of fastening the tread or riser to the carriage.

The whole process of shimming treads is slowed or voided if the center carriage is too high. When I cut my own carriages, I eliminate the possibility of the condition arising by overcutting both the treads and risers of the center carriage $\frac{1}{2}$ in. to $\frac{3}{8}$ in. By doing this I've eliminated the chance that the treads of the middle carriage will protrude past the line formed between the two outside carriages.

When it comes time to install the treads, I level and shim the two outside carriages. Then, it is simply a matter of gluing and screwing a 1½-in. cleat to the center carriage, which is brought into contact with the finish tread.

You can use the same process for the risers. By eliminating the center horse in the initial shimming process, leveling and shimming is transformed from a struggle into a dance.

Handrail jacks support rails in place

Securely holding a handrail during fitting and installation is infinitely easier with three handrail jacks.

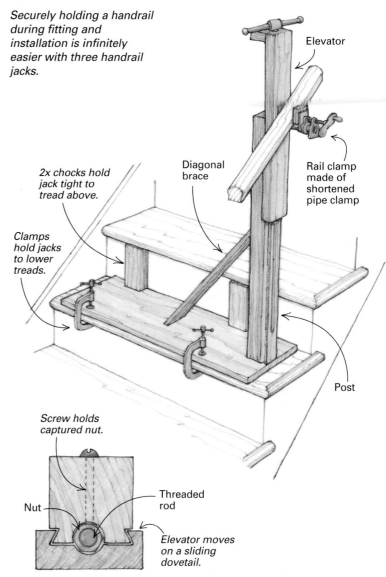

Elevator

2x chocks hold jack tight to tread above.

Diagonal brace

Rail clamp made of shortened pipe clamp

Clamps hold jacks to lower treads.

Post

Screw holds captured nut.

Nut

Threaded rod

Elevator moves on a sliding dovetail.

One day when I was scheduled to install a circular handrail, my helper didn't show up. To take his place, I made three rail jacks. The jacks are fairly complex, and they took a while to make. But they were well worth the effort: These rail-jack "helpers" are always on time in the morning, and they don't require a paycheck.

The jacks are made of a post and an elevator that ride along one another by means of a sliding dovetail. A piece of threaded rod, controlled by a T-handle at the top of the elevator, screws through a long nut (sold in hardware stores as a coupling for threaded rod) held captive in the stationary post. C-clamps at the base of the jack hold it to the finish tread. Short blocks of 2x brace the jack against the tread above. A diagonal brace, screwed to both the post and base of the jack, keeps the jack steady. Rail clamps are made of shortened pipe clamps.

When I installed the handrail, I used one jack at the top of the staircase, one at the middle of the staircase and one at the bottom of the staircase. When I am setting a circular rail, movement at any one of these three points is critical because any movement at one point has an effect on the other two points.

After using the jacks to ensure the rail is situated, marked, cut and fit correctly to the newels, I use a jack or two to steady the rail while boring holes for the spindles. Because of their unwieldiness, circular rails must be bored in situ, unlike straight rails, which can be bored upside down on sawhorses by means of a pitch gauge.

I attach a level vial to my drill when I bore spindle holes; this ensures the plumbness of the spindles.

Reinforcing the first step

Strengthening the rough stringer's first step with glued and nailed plywood gussets ensures the stringer won't break across the weak diagonal grain.

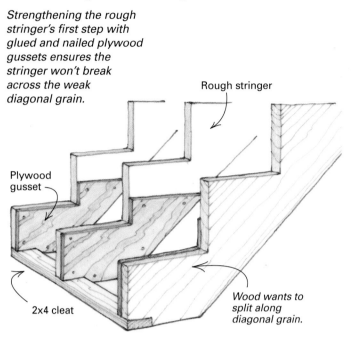

Rough stringer

Plywood gusset

2x4 cleat

Wood wants to split along diagonal grain.

Often, because the carriage at the first tread must be cut shorter than the unit rise to allow for tread thickness (usually 1 in.) and also notched for a 2x4 floor cleat, the bottom of the carriage is weakened. I always beef up the carriages by screwing and gluing a piece of plywood to the sides.

Covering newel-mounting bolts

Cutting a ¼-in. slice off the edge of the newel with a bandsaw and gluing it back on after running in the bolts both covers the holes and makes for a neat, finished look.

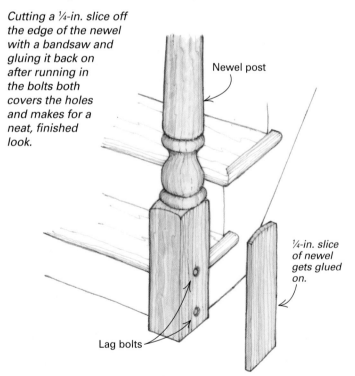

Newel post

¼-in. slice of newel gets glued on.

Lag bolts

Sometime the only way to fasten the newel post is to anchor it to the rough horses. (For an alternative method of attaching newels, see p. 57.) Often the bolt holes you have to drill are at an angle other than square to the face of the newel. Off-angle holes can be hard to plug, and when you do plug them, they usually don't look right. Instead of plugging holes, I saw ¼ in. off the face of the newel with a bandsaw, drill and install the bolts and then glue the piece over the holes. The bandsaw limits the loss of stock to about 1/32 in.—hardly noticeable—or you can plane off ¼ in. and make a new faceplate to glue over the holes.

Plumbing balusters

Marking your plumb line with a felt-tip pen makes a quick reference for plumbing successive sets of balusters.

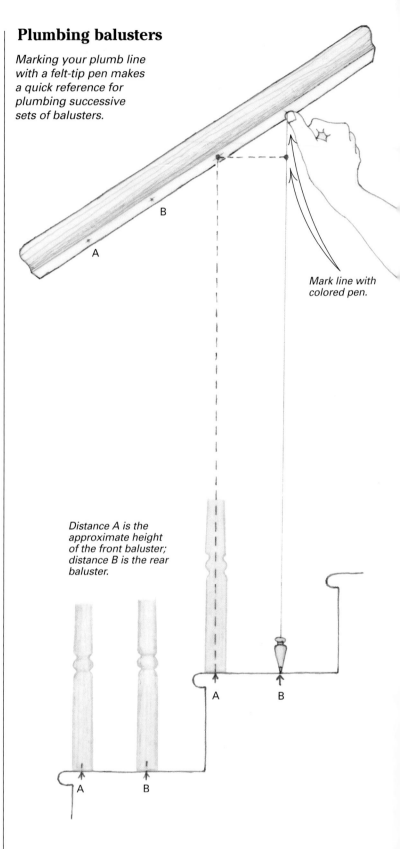

Mark line with colored pen.

Distance A is the approximate height of the front baluster; distance B is the rear baluster.

When laying out a handrail to drill holes for balusters, I use a plumb bob to get a vertical line from the marks I've laid out on the treads. I used to fiddle with the plumb bob's string length on each tread and wait for the bob to stop swinging. But now I make marks with a felt-tipped pen on the plumb bob's string—one at the height of the front baluster, one for the rear. When I go up to the next tread, all I have to do is hold the line on the mark for the particular baluster location and hold it on the rail. The approximate marks make a quick reference, and half the battle is over; I don't have to fumble with string length. Here's another hint: I've found that using braided string, as opposed to the more common twisted-strand string, helps to keep my plumb bob from spinning and swinging around when the bob is hanging free.

Baluster gauge

Baluster lengths can vary. On rails with a groove plowed on the underside, this gauge makes quick work of finding the proper length.

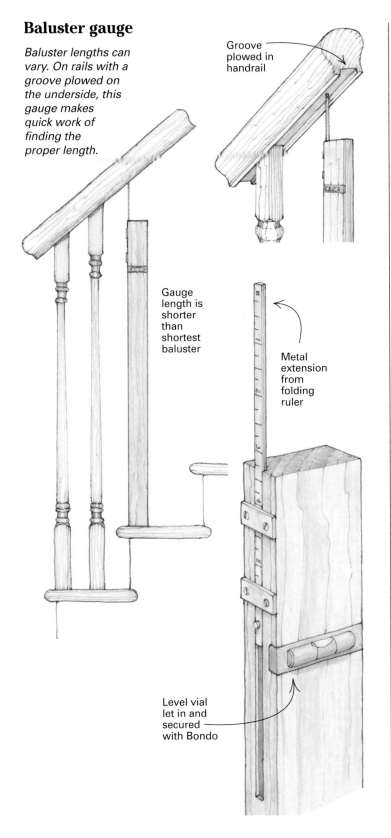

Groove plowed in handrail

Gauge length is shorter than shortest baluster

Metal extension from folding ruler

Level vial let in and secured with Bondo

Spindle-hole sizing gauge

Testing the fit of a tapered spindle in a block with different-size holes drilled in it determines the size hole to drill in the handrail.

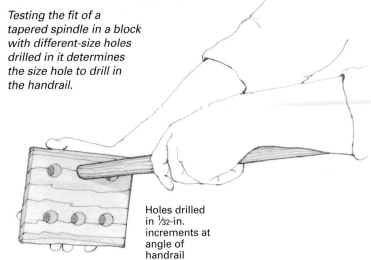

Holes drilled in 1/32-in. increments at angle of handrail

The diameter of the top, or thin end, of a tapered spindle can decrease as the spindle gets longer. Therefore, the hole drilled into the handrail for the back spindle on a tread can be smaller than the hole for the front baluster.

I take a block of wood and drill a series of different-diameter holes in it. The holes are drilled at the same angle that the spindles meet the handrail. After I cut a tapered spindle to length, I plug it into the sizing gauge to determine which size hole fits best, then I bore the hole in the rail. To save time, it's best to have two or three drill motors chucked up with the bits you'll most likely need.

Circular-rail center finder

Modifying a marking gauge eases the process of finding the center of a circular or elliptical handrail.

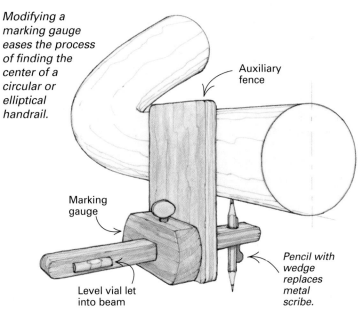

Auxiliary fence

Marking gauge

Level vial let into beam

Pencil with wedge replaces metal scribe.

Some staircases call for square-ended balusters that fit into a groove plowed into the underside of the handrail. In an ideal world all the balusters for a given position on each tread would be the same, and you could just go along and cut sets of short and long balusters. For various reasons, though, baluster lengths can vary as much as 3/16 in.

I made a baluster gauge that employs the sliding metal ruler taken from the end of a folding rule. The thin metal ruler is let into a piece of wood cut a couple of inches shorter than the shortest baluster. Small wood straps hold the ruler in place. A level vial let into the piece of wood makes plumbing easy. I secured the vial to the wood with Bondo. The same measurements could be had by trying to juggle a level and a folding rule, but the time saved using this shop-made gauge more than makes up for the time spent making it.

For finding the center on oval or round handrails that have been fit and either permanently or temporarily fastened, I modified a marking gauge by letting a level vial into the gauge's beam and replacing the metal scribe with a pencil held in place by a wood wedge. An auxiliary fence provides the additional height that is needed to compensate for the increased length of the pencil. By watching the level vial, I can keep the beam horizontal as I run the gauge down the length of the handrail, and I make a pencil line along its bottom center.

On a similar note, I've found that one of the most useful tools for both shop work and work in the field is a regular marking gauge with the metal scribe replaced with a mechanical pencil. □

Michael von Deckbar-Frabbiele is a former stairbuilder in New Orleans, La.

Plunge-Router Stairs

An adjustable mortising jig simplifies the construction of an open-riser stairway

by Bill Young

Leaving the risers out of a stairway can make a big difference in some houses. The open spaces between the treads allow for air circulation, and also a way for natural light to get into a sometimes claustrophobic corridor. Last summer, I built a pair of these riserless stairways for a three-story house, and it gave me two good problems to solve.

First, I wanted to mortise, or house, the treads into the stringer rather than notch them. I think housed treads look better, and this method also leaves the stringer intact, allowing for longer spans and a stairway with less bounce. The problem was how to cut the mortises quickly and accurately.

I knew all along that I would be making these cuts with my plunge router. This kind of router allows the bit to be lowered directly into the work, rather than let into it from the side. My first job was to figure out what kind of a jig would do the job precisely, yet still be easy to move along the stringer.

The second problem was to make the jig adjustable. Although the rises and runs for the two stairways were nearly equal, they had enough variation from floor to floor to require a different setup for each pair of stringers. My solution to both of these problems is the device I call the universal stair-stringer jig (photo facing page).

The jig consists of two basic parts: a rotating circular template and a cradle. A slot centered on the template's axis duplicates the cross section of the tread material, and guides the bit. Because I use a router bit that has a ball-bearing pilot mounted over the cutter—instead of under the cutter as is the usual case—I can size the cutout in the template to the exact profile of the stair tread. The cradle is a rectangular piece of plywood with a circular hole that is slightly larger than the template. Parallel fences on the bottom of the cradle position the jig on the stair stringer. So that it will be easy to move the jig along the stringer, I put the fences about 1/8 in. farther apart than the width of the stringer.

To use the jig, I lay the cradle over the stringer and I rotate the template until it aligns with the tread layout. This is where the universal part comes in. The jig can accommodate any tread layout, no matter what the rise or run of the stairs. When I'm satisfied with the alignment, I shim the fence against one side of the stringer (photo facing page). Then I use a bar clamp to lock the template inside the cradle, and to hold the jig in place as the stringer is routed.

Building the jig—I use 1/2-in. MDO plywood for my stair jigs. The letters stand for medium-density overlay. The overlay is a thin layer of plastic, which gives the plywood a smooth, hard finish. Around here a sheet of it costs less than $20. I've found MDO to be the perfect material for router jigs because it's strong and stable, it takes layout marks well and the router glides easily across its smooth finish.

I began building this jig by making the cradle from a 16-in. by 24-in. piece of MDO plywood (drawing, below left). I scribed a 13-in. dia. circle onto the rectangle, leaving three equal margins at one end of the rectangle. The wheel could be larger, but not much smaller. My treads were 11 in. wide, so a 13-in. template wheel gave me room for 1 in. of material at each end of the mortising slot. I drilled a 1/2-in. starter hole just touching the inside edge of the scribed circle, and I adjusted my electric jigsaw circle-cutting attachment to cut a 13-in. dia. hole. I began and ended the cut at the 1/2-in. starter hole.

I then cut a relief gap about 1/8 in. to 3/16 in. wide from the far end of the rectangle to the middle of the circle. The relief gap allows me simultaneously to clamp the template wheel at the correct angle, and to secure the cradle to the stair stringer.

Next, I attached 2x3 fences to either side of the cradle with 1-in. bugle-headed drywall screws. I used a piece of the stringer material to align the guides as I screwed them to the cradle.

It took some trial-and-error experimentation to get a correctly fitting template wheel. The tolerances are close, and the cutout left over from making the cradle was a bit too small to fit properly. I increased the radius of my jigsaw circle-

cutting attachment by 1/16 in., and got the right diameter wheel. When you finally achieve a correctly fitting template wheel, make several. Extras can be used for other tread dimensions.

To cut a precise slot in the template wheel I first placed the blank wheel inside the cradle and clamped the assembly to a suitable stringer board (photo next page, left). Then I scribed a centerline through the wheel and aligned it parallel to the long sides of the cradle. I used small nails to attach fences parallel to the centerline. The space between the parallel fences must be figured exactly since the fences will guide the router as it mills the template. This stair uses treads that are called 5/4 stock by the lumberyard, but they net out at about 1 1/16 in. thick. Since I was using a 3/4-in. bit to make my mortises, I had to gain 5/16 in. to arrive at the right mortise width. That meant that the space between the parallel fences had to be the width of the router baseplate plus 5/16 in.

To dimension the tread width, I scribed the outline of the tread on the template blank (photo next page, center) and cut to it freehand (photo next page, right).

Matching the treads—Most stock stair-tread material is rounded over on one edge to form the nose. On the oak tread stock we used, the factory-rounded edge could be used as it was, but we had to mill the unfinished edge with a 3/8-in. roundover bit to each side. This created an edge that fits perfectly into a mortise cut by a 3/4-in. dia. straight bit. This matching of the roundover bit to the straight mortising bit will occur again when the stair handrails are matched to the newel posts.

Laying out the stringers—My colleague Malcolm McDaniel came up with a slick way to determine tread layout for the router jig. The unit rise and run are drawn on the stringer using a carpenter's square in the usual way, but just for two treads—enough to measure directly with dividers the distance from nose to nose (drawing, below right). Next set your combination square

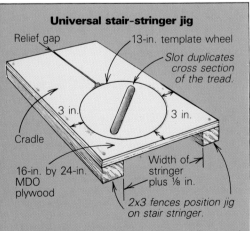

Universal stair-stringer jig

Relief gap
13-in. template wheel
Slot duplicates cross section of the tread.
3 in.
3 in.
Cradle
16-in. by 24-in. MDO plywood
Width of stringer plus 1/8 in.
2x3 fences position jig on stair stringer.

Young's adjustable mortising jig straddles a stair stringer (facing page) as he cuts a mortise with a plunge router. The heart of the jig (above) is a wheel held fast inside a housing, called a cradle. The wheel adjusts to any stair layout, and the entire assembly is secured to the stringer with a bar clamp and a pair of shim shingles.

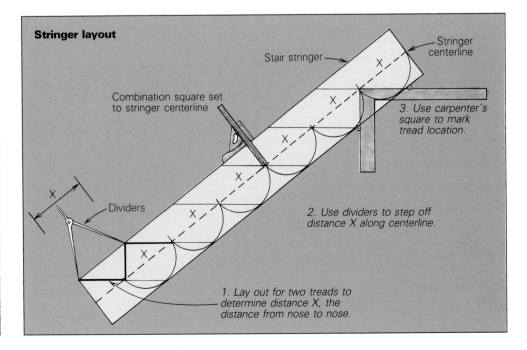

Stringer layout

Stringer centerline
Stair stringer
Combination square set to stringer centerline
Dividers
3. Use carpenter's square to mark tread location.
2. Use dividers to step off distance X along centerline.
X
1. Lay out for two treads to determine distance X, the distance from nose to nose.

Making the template. To cut the template, Young first draws a centerline through the wheel, left, and aligns it parallel with the long edges of the brake. Next he tacks a pair of parallel fences to the cradle to guide the router as it cuts the template. Young uses a section of the tread material to scribe its dimensions onto the wheel, center. Then he routs freehand to the scribe marks, right.

for half the width of the stair stringer. Holding the combination square against one edge of the stringer with one hand, swing the dividers with the other to step off the nose-to-nose increments down the middle of the stringer. At every point marked by the dividers, use the carpenter's square to mark the location of each tread.

To cut a mortise, place the jig onto the stringer and rotate the template wheel to the first run line. Now clamp the jig securely to the stringer and recheck the alignment. You may line up either side of the slot to the tread line so long as you are consistent for the entire stringer. To avoid confusion, I make a mark on the side of the line that I intend to mortise. Once I'm satisfied with the alignment, I carefully plunge my mortising bit through the slot in the template and into the workpiece.

I cut ¾-in. deep mortises for this stair. Each one needed two passes with the plunge router—a ⅜-in. depth setting followed by a ¾-in. setting. Because you will need tail pieces to clamp the jig, the stringer should not be cut to length until the last mortise is routed.

Assembling the treads and stringers— After the treads are rounded over and cut to length, they can be let into one of the stringers that has been laid on the floor, mortise side up. Our fit was snug enough to require a gentle assist with a hammer, interceded, of course, by a piece of scrap. We placed the second stringer atop the standing treads and tapped it home, starting at one end and working little by little to the other end. We didn't use any glue because there is little lateral force on the stair. Instead, we predrilled the stringers and counterbored at each mortise for three 2¼-in. bugle-headed screws—primarily to keep the stair together during installation. On the exposed stringer, the screw holes are filled with hardwood plugs.

Newel posts— We made our newel posts from clear, kiln-dried Douglas fir 4x4 stock, reduced to 3 in. by 3 in. for a slimmer profile. Each post is secured by a ⅝-in. threaded steel rod, which

is screwed into a steel flange mounted to the floor (drawing, facing page, bottom left). I began making the newel posts by sinking a deep counterbore into one end of each 4x4. I used a 2-in. Planetor bit (Rule Industries, Planetor Div., Cape Ann Industrial Park, Gloucester, Mass. 01930) for this operation because it cuts well in end grain. I chucked the bit in a floor-standing drill press and mounted the 4x4 stock on a stop-jig that kept it vertical and resisted the bit's torque. I used the same arrangement to drill a shallower, smaller-diameter counterbore in the opposite end for the floor-mounted flange nut.

The next step was to rip the 4x4s down the middle with a bandsaw and surface the sawn sides smooth on a jointer. Then I routed a groove down the center of each half with a ¾-in. core-box bit, connecting the counterbores at each end. During all of this milling, I was careful to keep track of the matching halves. Milling completed, I glued and clamped each pair back together. After removing the clamps, I brought the still oversized newel posts down to their final dimension of 3 in. by 3 in. on a planer, removing equal waste on every side so that the rod hole remained centered in the post. The result of this procedure was a solid-looking newel post, almost indistinguishable from a post that hadn't been cut in half and reassembled.

I had the mounting plates made up at a local sheet-metal shop from ³⁄₁₆-in. steel, with a ⅝-in. nut welded to the center of each one. The plates are held in place by four 3-in. #10 screws, driven into doubled joists at each landing. The finish floor is covered with tile, so I didn't bother to mortise the plates into the floor.

Handrails and balusters— We used kiln-dried Douglas fir for the handrails and balusters. The handrails are 2x4s, rounded over on all edges. (Check building codes for design requirements.) The balusters are 1x2s. To house the balusters, I plowed a ¾-in. wide groove in the underside of all the handrails, and a corresponding groove on the top edges of the stringers. The balusters are positioned and secured by 1x spacers.

Mortising the newel posts— Each newel post is mortised to receive both the handrail and stringer. I cut both these mortises with the same kind of template that I used on the stringers, but I used a 1-in. dia. bit to match the ½-in. round-over on the handrails. Before I made any cuts, I temporarily installed the newel posts and made sure that they were plumb. This let me accurately measure the length of the handrails, and determine the plumb-cut face of both the handrails and the stringers.

The mortises in the newel posts are ½ in. deep, and the handrails are glued in place. The sharp angles at the end of the handrails and stringers have to be squared off so the assembly will fit properly (drawing facing page, top right).

Assembly— After threading the steel rods into their floor-mounted nuts, we lowered each newel post over its rod. Then we glued the stringers and handrails into their mortises, and cinched down the posts with a deep socket wrench. Next, we inserted the balusters into the slots on the top of the stringer and the bottom of the rail, alternating spacers and balusters and gluing the entire assembly. We topped the newel posts with an oak cap.

After cleaning up excess glue and sanding, we varnished the entire stairway with four coats of satin marine spar varnish. The result is a clean, spare and surprisingly strong stairway (photo facing page). The universal stair-stringer jig did what we needed it to do. It guided the router bit with precision, and it reduced the time spent preparing each stringer to less than is spent cutting a conventional stringer. □

Bill Young is a contractor in the San Francisco Bay Area.

The finished stair (facing page) is a light yet strong assembly of mortised parts that lets light and air into the stairwell. Except for the newel posts, which have been planed to 3x3s, all of the stair parts are made of dimension boards straight from the lumberyard.

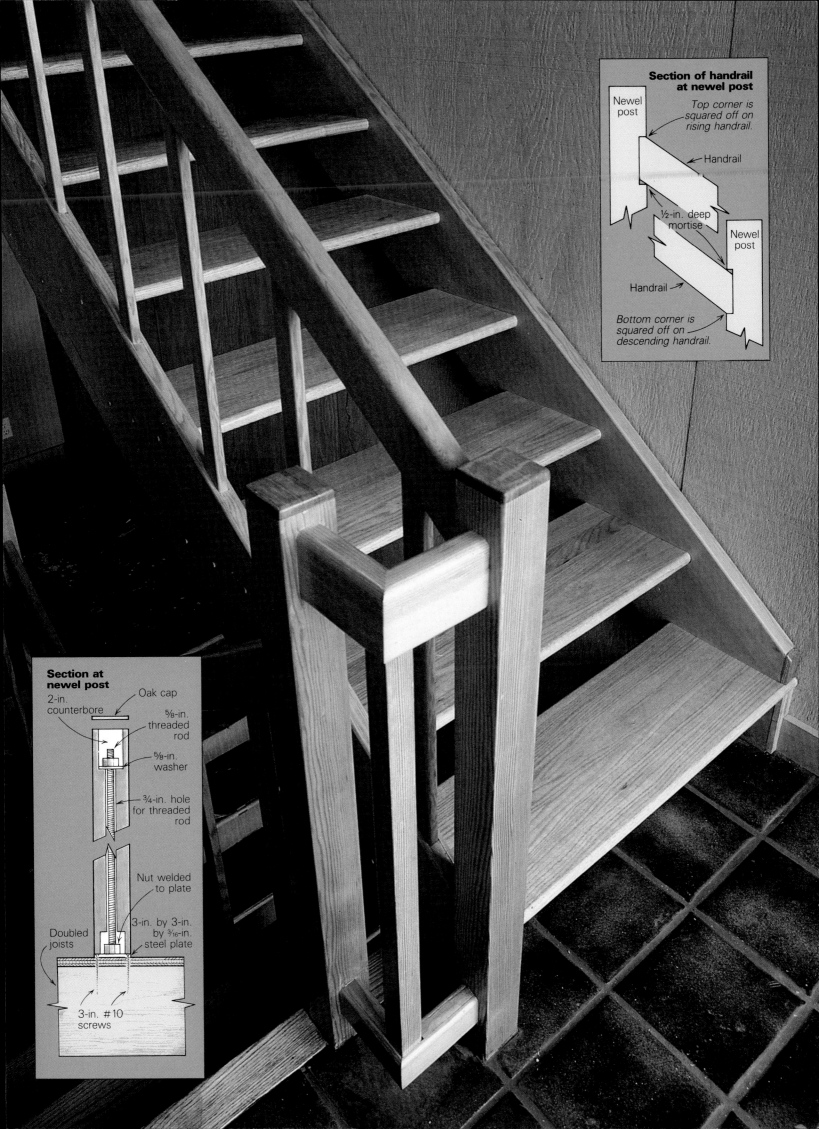

Section of handrail at newel post

Newel post

Top corner is squared off on rising handrail.

Handrail

½-in. deep mortise

Newel post

Handrail

Bottom corner is squared off on descending handrail.

Section at newel post

2-in. counterbore

Oak cap

⅝-in. threaded rod

⅝-in. washer

¾-in. hole for threaded rod

Nut welded to plate

3-in. by 3-in. by ³⁄₁₆-in. steel plate

Doubled joists

3-in. #10 screws

Rising from what once was a bedroom, this new stair leads to a second-story addition on a turn-of-the-century Arts and Crafts home. Builder Alexander Brennen built the stair on site, using Douglas fir newel posts, oak treads and risers to match the flooring; railings and balusters to correspond with the original trim elements.

A Site-Built Stair

Using readily available materials and on-site carpentry techniques to build a tasteful staircase

by Alexander Brennen

It's easy to see why carpenters are attracted to stairs. Along with framing a roof, building a staircase is one of the most challenging geometrical tasks in building a house. And once the variables of rise, run, headroom, railing and landing configurations have been resolved, the carpenter assigned to build the stair can look forward to airing out some of the finish-carpentry tools that have been languishing in the corners of his toolbox.

Architects can also fall victim to the spell of a well-turned stair, and they often design elaborate stairs—no doubt at the request of their clients. Unfortunately, complicated stairs are frequently beyond the budget. Sometimes they must even be built off site and reassembled in place. A pleasing staircase can, however, be built using standard on-site carpentry methods. This article is about such a stair (photo above). It connects the ground floor of

a turn-of-the-century Arts and Crafts-style house to a new upstairs addition, designed by architect Glen Jarvis.

From the ground floor up, the stair has two primary flights connected by a landing. Another landing on the second floor leads to a short flight with only three risers. Glen's original design for the stair detailed a traditional oak balustrade assembled from manufactured parts. But as we got further into calculating the costs of the remodel, it became clear that the money wasn't going to be available for expensive stair parts and their fastidious fitting. Jarvis and I met with the owners of the house, Morris and Regina Beatus, mulled over our options and decided to build a simpler stair in keeping with the original house. The treads and risers would still be oak, for its durability and to match the oak floor in the upper and lower halls. The railing would be of clear Douglas fir, which would match the door and window trim.

Horses on rake walls—Once we had the second floor framed and the roof in place, we calculated the rise of the stair by measuring the distance from the existing oak strip floor in the lower hall to the top of the subfloor in the upper hall. To this number we added $5/16$ in. to account for the thickness of the oak strip flooring that would cover the upper-hall floor. We then divided this number by the number of risers to establish the riser height. For this stair, the rise ended up at $7\frac{3}{4}$ in. and the run at 10 in.

The landing between the two primary flights of stairs is 6 ft. wide, which is the minimum width allowable by our code to accommodate our 3-ft. wide treads—also a code minimum. Once we knew our riser heights, we began our stair framing by building the lower landing first. There was nothing tricky about this part of the project because we used standard framing lumber and conventional stud-wall construction techniques to build the stair's superstructure (photo below right). The stair horses (also called carriages, stair stringers or stair jacks in some parts of the country) were cut from 2x12s. On the open side they bear on 2x4 rake walls. On the wall side they are affixed with 16d nails to ledgers that are anchored to the stud walls (drawing, p. 29). Drywall backing blocks made from 2x10 stock fill the stud bays adjacent to the 2x6 ledgers. Once we had the landings and horses in place, we installed some temporary treads and didn't work further on the stair until the drywall work was complete.

Newel posts and treads—The newel posts are 4x4s made of clear, dry Douglas fir. I notched the bottoms of each post to fit beside the stair horse (section drawing above) or the landing rim joist. After plumbing each newel post so that it was parallel with the rise of the stair, I glued each one to the framing with PL 400 subfloor adhesive (ChemRex Inc., 7711 Computer Ave., Minneapolis, Minn. 55435; 612-835-3434) and secured each with a pair of $3/8$-in. machine bolts. In places where I couldn't run the bolts, I used a half dozen 4-in. galvanized drywall screws driven from different angles to lock the post firmly in place.

I had planned to use the laminated oak-tread stock available from our regular supplier, but after learning that I could buy enough 4/4 red oak to make the treads, risers and skirtboards (sometimes called stringboards) for the same amount of money it would cost

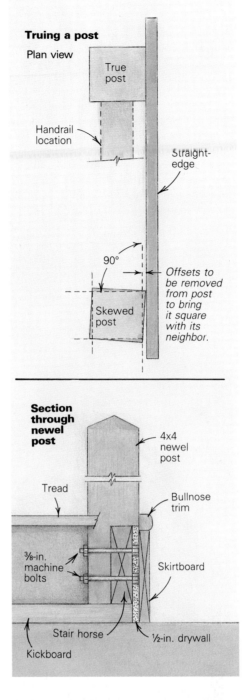

Truing a post

Plan view

True post

Handrail location

Straight-edge

90°

Skewed post

Offsets to be removed from post to bring it square with its neighbor.

Section through newel post

4x4 newel post

Tread

Bullnose trim

$3/8$-in. machine bolts

Skirtboard

Stair horse

½-in. drywall

Kickboard

Two straight flights of stairs engage the first landing, which is built using standard stud-wall framing techniques and materials.

to buy just the premade treads, I decided to make my own. To trim the treads, I would use $1\frac{1}{16}$-in. sq. bullnose trim on the front and open side. Using a separate piece for the bullnose trim would allow me to wrap the treads around the newel posts easily, and to hide the end grain of the treads.

In our area, the most economical red oak is available from a local supplier who has a pile of it in random widths and random lengths. The boards are planed on two sides and jointed on one edge only. In any given stack, most pieces are between 4 in. and 6 in. wide and about 8 ft. long. I found a few wide pieces in the pile that could be used for skirtboards and risers. Then I picked out boards that were wide enough to allow me to rip them into 5-in. boards.

Working in my shop, I cut the boards in half and glued their edges together to make 10-in. wide, 4-ft. long planks. This would be the tread stock. Because the planks were cut from the same board, their grain matched well and their figure had more character than the premade treads, which are often glued up from 2-in. wide pieces.

After gluing up the treads I ran all the oak through a 12-in. thickness planer/jointer. After a couple of passes to put their faces in plane, I ended up with pieces $3/4$ in. thick. I then jointed their edges and ripped all the various parts to size on the table saw.

Job-site assembly—Once the drywall crew was through making and cleaning up their traditional mess, I brought all my oak stair parts to the site. At last I could yank the temporary treads off the stair horses and get down to business. I installed the skirtboards first. On the wall side they tucked against the drywall in the gap between the horses and the 2x6 ledger. After cutting the outside skirtboards to fit at top and bottom, I clamped them in place and marked them for the tread and riser cuts. The risers and outside skirtboards met at a 45° miter (bottom left photo, following page), so the layout mark represented the backside of the cut.

Because there are two outside skirtboards running in opposite directions, I needed to make miter cuts from opposite sides of the boards. Fortunately, my worm-drive Skilsaw tilts one way, while my Porter-Cable worm-drive trim saw tilts the other way. I used sharp combination blades to make the cuts, and clamped straightedges to the skirtboards to guide them.

The skirtboards are affixed to the framing with 2-in. galvanized drywall screws. All the screw holes are plugged. To bore each screw hole and plug hole at the same time, I used a tool with a 6-in. long, $1/8$-in. dia. bit and a movable #6 countersink, which carves a $3/8$-in. dia. hole for a plug (W. L. Fuller, Inc., P. O. Box 8767, 7 Cypress Street, Warwick, R. I. 02888-0767; 401-467-2900).

I installed the treads and risers from the bottom up, fitting the first two risers and then the first tread. Each one had to be scribed to the skirtboard, and that's where having 4-ft.

stock for 3-ft. treads came in handy. I could afford to be finicky about the fit, knowing I had some extra stock to let me whittle away at one end if need be. I marked my scribe lines with a sharp utility knife, and cut to the line with a trim saw set at a 2° bevel to give me a slight back-cut. Then I used a block plane to make minute adjustments. A sharp block plane is a necessity on a job like this—I kept mine busy on every phase of the finish work.

Alternating riser, then tread as I worked my way up the stair, I fastened the risers to the outer skirtboards and the treads to the horses with 8d galvanized finish nails and aliphatic resin glue. From the back of each riser I drove 2-in. galvanized drywall screws, 9 in. o. c., into predrilled holes in the adjacent treads, and used the same screws to anchor the leading edges of the treads to the risers below. They too are glued. I anchored the back edge of the top tread to a ledger that is screwed and glued to the framing. Where a riser abutted a newel post, I checked to make sure that the post was plumb in both directions. If it wasn't, I leaned on the post a little, and drove screws through the riser into the post to bring it plumb. Once I had all the treads and risers in place, I glued and nailed bullnose trim to all the treads and around both landings.

I plugged the countersunk screw holes with oak plugs, and pared them flush with a chisel. Then I used a hand scraper to remove the excess glue and smooth out the edges between the treads and the bullnose trim. Using a scraper sounds fancy, but once you get the hang of sharpening it, the scraper is a very fast tool—especially when working with oak. Before bringing on the flooring subcontractor, I filled all the nail holes and sanded the entire stair.

After beveling the tops of the posts with a saw, Brennen dresses the cuts with a sharp block plane (top photo). To avoid splitting the ends of the balusters, Brennen predrills all nail holes while holding the piece in place (middle photo). The risers and outer skirtboards are cross-nailed and glued where they meet at mitered corners (bottom photo). Here, a tread notched to accommodate a landing newel post is lowered onto a fresh line of aliphatic resin glue. The tread's ends will be nailed to the horses, while its front and back edges will be screwed to the adjacent risers. The raw edge of the tread will be covered with oak bullnose trim.

Precise bevels. Using a clamped-on jig to guide his worm-drive saw, the author begins the first of four bevel cuts that will result in shallow, pyramidal tops for the newel posts. The cuts here were made with the saw set at 22.5°.

Railing—I returned to the project to work on the railing and balusters after the two landings and the upstairs hall had been laid with oak strip flooring, and everything—stair included—had been finished with a light stain and three coats of polyurethane.

Simplicity remained our watchword, as we decided to use a stock mushroom-type handrail affixed to the top of a 2-in. by 2½-in. piece of Douglas fir (middle left photo, facing page) to give the railing some mass. But before I could install the railing between the posts, I needed to cut their tops to length.

I used my worm-drive saw along with a clamped-on guide to bevel the tops of the posts to a shallow peak (large photo, facing page). For this bevel, I set the base of the saw to make a 22.5° cut.

Next I used a long straightedge, held against adjacent posts, to see if they were square to one another. Nope. As shown in the top drawing on p. 27, which is exaggerated for clarity, the posts were skewed in relation to one an-

other. This didn't make any difference structurally, but it made it difficult to fit a railing precisely between them. To correct the situation, I marked the top of the skewed post and used my block plane (what else) to taper the post so as to put it in plane with its mate. Then I squared off this line to make each face of the post match its neighbor. This doctoring doesn't extend all the way down the post. It's actually a slight corkscrew that isn't noticeable.

I installed the lower portion of the railing first. I clamped it to both posts at the correct height (34 inches in this case) and used my knife to scribe the angle of cut directly on the rail. I made these cuts with a 14-in. power miter box, and left the setting the same for the handrail cuts. The lower portion of the rail is anchored to the newel posts with a couple of 3-in. galvanized drywall screws at each intersection. The screw holes were covered by the mushroom-cap portion of the railing, which I affixed to the bottom rail with glue and screws

driven from below so that no fastener holes show on the topside of the rail.

The 2x2 balusters are spaced evenly on 5-in. centers. Two balusters fall on each tread, and they are toenailed to the treads from opposite directions and to the underside of the handrail with 6d finish nails driven into predrilled holes. I laid out their positions on the treads and railing with light pencil marks, and cut all the balusters at once. There are two lengths, and I cut each one slightly long (½4 in.) to allow myself some adjustment. After a test fit, I nailed each one in place, securing their tops first because the angle of the rail held them steady.

By the time I tacked the last baluster in place, Morris Beatus was ready to take over. He filled the nail holes, sanded the unfinished surfaces and then finished them with three coats of Watco oil. □

Alexander Brennen is a partner in Zanderbuilt Construction in Berkeley, California.

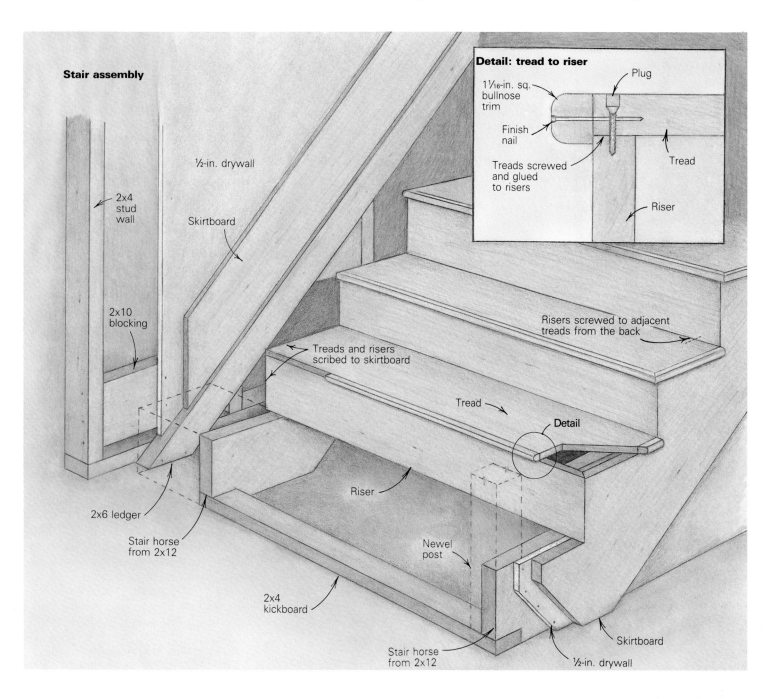

Stair assembly

½-in. drywall

Skirtboard

2x4 stud wall

2x10 blocking

Treads and risers scribed to skirtboard

2x6 ledger

Stair horse from 2x12

2x4 kickboard

Riser

Tread

Newel post

Stair horse from 2x12

½-in. drywall

Skirtboard

Risers screwed to adjacent treads from the back

Detail

Detail: tread to riser

Plug

1¹⁄₁₆-in. sq. bullnose trim

Finish nail

Treads screwed and glued to risers

Tread

Riser

Storage Stair

An alternative to conventional framing takes advantage of normally wasted space

by Tom Bender

Astairway that my wife Lane and I saw in Japan about 10 years ago etched itself in our memories. Instead of being built on notched stringers, it was made up of finely joined chests and boxes of graduated heights. In stepping from the top of one box to the top of another, you arrived at the second floor. Needless to say, when the time came to build a stair in our own house recently, the Japanese stairway was still in our minds. Our stair (photo, left) contains four different kinds of storage units: drawers, cupboards, a roll-out toybox and a book alcove. There's even a drawer under the lowest tread that pulls out for storing slippers and socks. Considering all the stuff that we're able to store in this normally wasted space, building the storage stair wasn't that much work. It was more work, of course, than for a conventional stairway, but no more than for a storage wall.

Design and layout—Our visions of fine Japanese joinery had to adjust to our own reality. We had an old Sears table saw, a skillsaw, a cheap belt sander, and a random assortment of hand tools. Our cabinet shop was a corner of the living room with the rug rolled up. We couldn't afford good hardwood, and didn't want a house too precious for people to live in. So we had to make do with the pile of construction-grade fir stacked in our living room, some used shiplap, and someone who had never built a stair before.

The structural system I worked out for the stair is a framework supported by four vertical partitions that serve as dividers between drawers or cabinets. These are connected by a grid of 2-in. by 1½-in. face-frame pieces that form a base for treads and act as drawer and cabinet bottoms.

I built all the partitions first. Each one is a different height, but they are all constructed the same way. An identical pair of 2x4 uprights was grooved to accept a ½-in. thick plywood panel and notched at the top for the 2x10 riser that is an integral part of each partition. I drew up a checklist of all the framing members and hung it up by the table saw, then crosscut long stock to rough length with my skillsaw and brought it to the table saw for finish cutting. The 2x4s for each partition were cut in pairs, since they would run clear

Tom Bender is an architect and builder. He lives in Nehalem, Ore. Photo by the author.

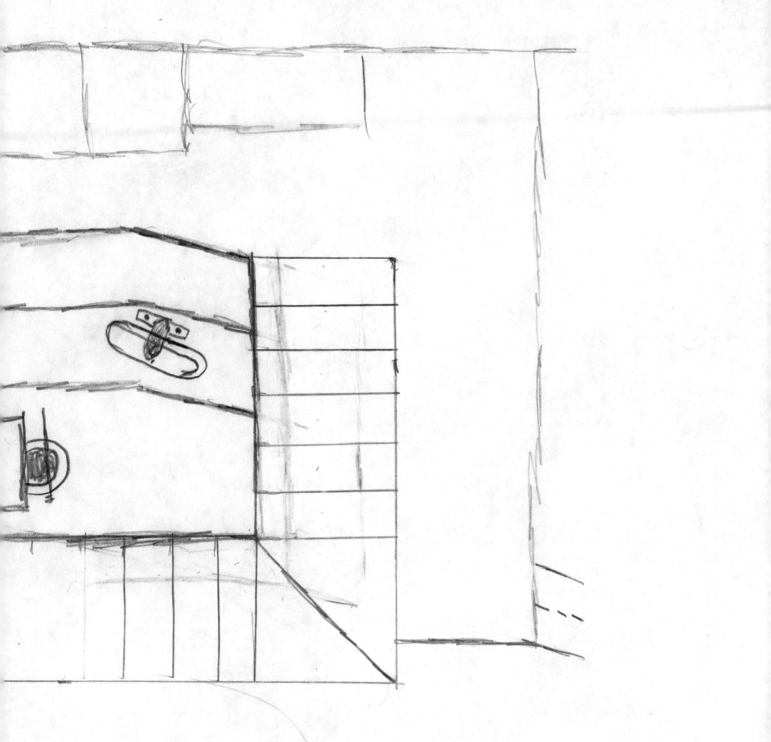

to the floor. I chiseled shallow (⅛-in.) mortises in the 2x4s to accept 1½-in. by 2-in. horizontal face-frame pieces that would connect adjoining partitions and also support the shelves, drawers and treads. These pieces would be joined to the 2x4s with glue and dowels, so I made a cardboard template and stuck a nail through it to mark dowel centers.

Assembly—Putting the whole thing together would have been easy for an octopus with a truckload of clamps. But I was armed with only two bar clamps, a few C-clamps and two hands. Partitions were assembled first. While they dried, I located their positions on the floor, cut all the horizontal pieces to size and checked dowel alignment on each joint. For the final assembly, I put up the tallest partition first, at the high end of the stairs. After gluing and doweling all eight of the horizontal pieces to this partition, I snugged the unit into place with the aid of my faithful bar clamps. After checking for level, I toenailed it to the floor. The remaining three partitions went up in the same way. Each one was clamped to its steady, previously installed neighbor before being nailed to the floor.

We let the glue in the completed framework dry overnight, and began on the stair's ten treads. As all our clamps would be needed to hold a single tread in position until the glue dried, I decided to nail-clamp the joints to hold them while the glue set. I drilled undersized holes, then nailed through the back of the upper riser with 16d box nails, and up through the side frame with 10d finishing nails. Then the bar and C-clamps could be removed and used to set the next tread. After a touch-up sanding, the frame and treads were finished with two coats of boiled linseed oil.

Doors, drawers and shelves—With a usable stairway complete, installing the shelves, drawer glides and stops, and building the doors and drawers could go more slowly. To match the walls of the living room and entry, I made all the doors with the same recycled shiplap, using horizontal battens on the inside face of each door to hold the shiplap boards together. Drawers were made up from ½-in. plywood, with Masonite bottoms. Shiplap was glued up to make each drawer face. For the handrail, I used a long piece of driftwood that we had found on the beach. I cut it to length, then notched it to fit against a 3x3 post that I attached to the stair frame with lag screws.

A translucent screen—Next, we needed to enclose the living-room side of the stairs to make the stairway safe for kids and also to keep heat downstairs. We didn't want the stair and entry to be dark, so a translucent wall seemed like a good idea. Everything pointed toward a Japanese-type screen. The modified *shoji* design we used was easy and inexpensive to build.

A frame for the translucent wall was made from a few 2x4s ripped into strips. I installed ¾-in. by 1½-in. strips vertically, attaching them to the edges of the treads and to the

Varied storage, step by step

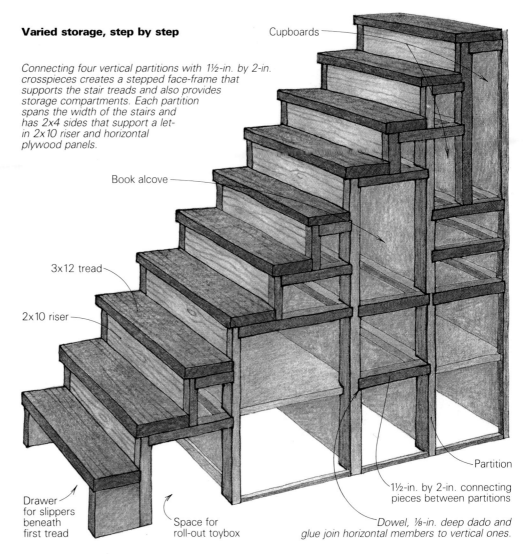

Connecting four vertical partitions with 1½-in. by 2-in. crosspieces creates a stepped face-frame that supports the stair treads and also provides storage compartments. Each partition spans the width of the stairs and has 2x4 sides that support a let-in 2x10 riser and horizontal plywood panels.

Cupboards

Book alcove

3x12 tread

2x10 riser

Drawer for slippers beneath first tread

Space for roll-out toybox

Partition

1½-in. by 2-in. connecting pieces between partitions

Dowel, ⅛-in. deep dado and glue join horizontal members to vertical ones.

edge of the floor above, which made them strong enough to prevent anyone from falling through. Smaller ⅜-in. by 1-in. strips were used horizontally, spaced half a riser apart, to stiffen the frame and support the paper covering. I notched all the pieces to a depth of ½ in. at all joints, gang-cutting with a dado blade on my circular saw to ensure correct alignment.

After a trial assembly to make sure everything lined up properly, the notches were then dabbed with glue, and the strips snugged into place with the aid of C-clamps. The inevitable curvature of the long, thin strips of wood caused some concern at first, but any unevenness was pulled back into line as the grid grew. The frame was sealed with a coat of linseed oil, except on the surfaces where the paper would be attached, and allowed to dry well, so that it wouldn't bleed onto the rice paper that would be glued onto it.

To apply the rice paper, I used a paste made of white flour and water. This makes it easy to remove the paper if any sections get damaged. The rice paper, which is commonly used for *sumi* ink drawing, is available in 8½-in. and 11-in. wide rolls from most art or oriental-import stores. We used the larger size, applying it vertically in sheets no longer than about 2½ ft. We found that if we used longer lengths, the glue we painted on the lattice framework would dry before the paper was in place. We worked from one end of the wall to the other, doing one vertical section at a time. A little

thumb pressure stretched out any wrinkles or sags, and we were able to keep the paper taut against the frame without much trouble. After waiting a few minutes for the glue to dry, I trimmed off the excess paper with a single-edge razor blade. Then we applied the next strip of paper, lapping the first by the thickness of the wood frame.

A door with rice-paper panels at the stairway finished off the job. Double-swinging spring hinges on the door allow us to go in and out with the ever-present armfuls of children, groceries, laundry or firewood, without losing too much heat or having to stop and close the door behind us. A wide mid-rail on the door acts as a bumper bar, allowing us to open the door even with our arms full.

The finished stair and storage space have met almost all our expectations. The rice paper creates a soft light that accentuates the rich color of the steps. And at night, an almost magic silhouette of the gnarled, curving handrail is cast against the precise and delicate rectangles of the grid.

We had feared that the rice paper wouldn't last long with small children around, but we have all learned to live with its delicacy pretty well. Occasionally holes do get punched through it, but they are easily patched with a rice-paper snowflake or butterfly. The storage works out beautifully, with sweaters, blankets and such available from either the entry or the living-room side of the stair. □

Disappearing Attic Stairways

Folding stairs are the most common, but sliding stairs are easier to climb

by William T. Cox

Sliding stairway. **Disappearing stairways are concealed by a spring-loaded ceiling door. Here, the author walks up a sliding stairway made by Bessler with an angle of incline close to that of a permanent stairway.**

Folding stairway. **Ladderlike sections are hinged like an accordion to the ceiling-mounted door. On this model, made by American Stairways, the treads are painted with bright-colored, rubberized paint.**

When I was young, and my mother wanted something out of the attic, she would push me up a stepladder and through a little access hole in the ceiling; it was a scary adventure for an 8-year-old, climbing up into a dark, cavelike hole where I thought unknown creatures waited to devour me. What we needed was a disappearing stairway.

Disappearing stairways are available in several styles. All of these stairways have a ceiling-mounted trap door on which the stairway either folds or slides. Nearly all are made of southern yellow pine, although there are a few aluminum disappearing stairways. There are a few commercial models made of aluminum or steel, but this article will concentrate on residential models. Disappearing stairways are not considered to be ladders or staircases, and they do not conform to the codes or the standards of either. Disappearing stairways have their own standards to which they must conform.

Similar to ladders, disappearing stairways have plenty of labels and warnings to read. On all disappearing stairways there are warnings about weight limits because, inevitably, homeowners fall down stairs while trying to carry too much weight into the attic. Also, labels tell the user to tighten the nuts and bolts of the stairway.

In one stairway manufacturer's literature, the word "safer" was used to describe the fluorescent orange paint used on the stairway's treads. But "safer" was replaced by "high visibility" because one homeowner wore off the paint, slipped and fell. She sued both the manufacturer and the builder because, she claimed, the treads became unsafe to use.

Stairway companies are constantly testing, upgrading and improving their products to give the consumer the best, safest and longest-lasting disappearing stairway possible. And with good reason—over a million units were produced in the United States last year.

Folding stairways—The most popular style of disappearing stairways, folding stairways consist of three ladderlike sections that are hinged together, accordion style. The three sections are attached to a hinged, ceiling-mounted door similar to a trap door. The door and the attached ladderlike sections are held closed to the ceiling by springs on both sides. When you want to access a folding stairway, you pull a cord that is attached to the door and lower the door from the ceiling. The door swings down on a piano hinge. You then grab the two bottom sections of the stairway and pull them toward you, unfolding them (right photo, above). When the two bottom section are completely unfolded, all three sections butt together at their ends, giving strength and stability to the stairway.

Folding attic stairs are measured by the rough opening they occupy and by the floor-to-ceiling height they will service. The smallest folding stairways are 22 in. wide, and they are made to fit between joists 2 ft. o. c. These narrow stairways are available in models that will service a ceiling height as short as 7 ft., and there are others that can go as high as 10 ft. 3 in. Keep in mind that a stairway's rough-opening width is appreciably more than the actual width of the ladderlike sections. Because of the attendant jambs, springs and mounting hardware necessary to operate the stairway, the actual width of the ladderlike section is a lot less than the rough opening. A stairway with a rough opening of 22½ in. is going to have a tread about 13 in. wide.

Folding stairways are rated according to weight capacities; the lightest-rated ones will handle 250 lb., and most of the others have a recommended weight capacity of 300 lb. It is interesting to note that American Stairways, Inc., says in its product literature that you are not supposed to carry anything up or down its stairways. Only an unladen person is supposed to climb up or down. This all sounds somewhat ridiculous to me; it's not as if someone is going to go up into their attic crawlspace simply to spend a little quality time. The reason why people install disappearing stairways is that they can carry stuff up or down from the attic—Christmas ornaments, baby clothes. However, I tell customers not to carry stuff up the folding stairway. You should have someone hand it up to you. You cannot climb a folding stairway with something in your hands. It's way too steep. I suppose the disclaimer keeps American Stairways out of court if somebody falls down one of the stairways. Also, all folding stairways are for residential use only; a restaurant owner once asked me to install a folding stairway so that he could access a storage area above the kitchen, and I had to refuse.

The smallest folding stairway costs around $75, and the largest, the A-series aluminum folding stairway made by Werner (see sidebar, p. 35) costs around $211. It fits a rough opening of 2 ft. 1½ in. by 4 ft. 6 in. and accommodates a ceiling height of up to 10 ft. 3 in.

Installing folding stairways—Aside from the finish trim, folding stairways come out of the box as a complete, assembled unit. (Other types of stairways require some assembly.) Because most of the installations I do are retrofits into existing buildings, the first thing I must do is cut a hole in the drywall. If possible, I try to mount the stairway alongside an existing joist; this saves some framing work if the stairway is bigger than the space between two joists. Cutting the drywall is not a close-tolerance operation because (within reason) the finish trim will cover any ragged edges. If I have to head off a ceiling joist, I use standard carpentry practices.

Here's a time-saver I came upon after installing quite a few stairways. I've found that it's much easier to cut and fit (but don't nail) the finish trim while the stairway is sitting on the floor in front of me rather than on the ceiling. Leave ⅛ in. between the edge of the door and the jamb. Make sure you mark the location of all four pieces. Once the stairway is installed, you just nail the pieces in place.

Before installing the stairway, I screw two temporary ledgers to the ceiling that project ¾ in. in-

Support during installation. Ledgers screwed to the ceiling provide temporary support for the stairway while the author shims and screws the frame to the rough opening.

An extra screw for insurance. A third mounting screw in a folding stairway's piano hinge strengthens the installation. The author drills through the hinge and the jamb and into the rough framing. The large spring at the top of the photo is one of a pair that holds the stairway and its trap door closed to the ceiling.

to the rough opening. The ledgers provide a shelf for the stairway's wood frame once I've lifted the unit into the rough opening (top photo, this page). When attaching the ledgers, I make sure they are parallel and that they only stick into the opening ¾ in. Any farther than that, and they might not allow the door to swing open on its piano hinge. Using screws instead of nails to attach the ledger makes it possible to adjust them in case I somehow miscalculate; it also makes them easier to remove when the time comes.

Although some manufacturers warn against it, I usually remove the bottom two ladderlike sections of the stairway before carrying it up the stepladder. Most often I work alone, and some of the stairways are pretty heavy to lift by myself. A 30-in. by 54-in. stairway made by American Stairways, Inc., weighs 92 lb.

With the ledgers in place, I lift the stairway into the rough opening and set it on the ledgers. Next I carefully open the door fully and center the

jamb in the rough opening. Now it's just a matter of shimming the sides of the frame and fastening them to the framing. (Once the unit is installed, I reattach the sections and tighten the nuts and bolts on the hinges with a screwdriver and a socket wrench.)

Most instructions call for nailing the frame in place, but I like to use screws because they are more adjustable than nails, and they are also easier to remove if needed. I start at the hinge end of the stairway jamb. Most folding stairways have a hole drilled at both ends of the piano hinge to screw the hinge into the framing. I always drill another hole through the hinge and sink a third screw (bottom photo, this page). I use #10, 3-in. pan-head screws. Adding a third screw can't hurt, and it only takes an extra minute or two.

Instructions call for screwing or nailing into the framing on both sides of the stairway through two of the holes drilled in the arm plate, which is the metal plate to which the door arms are attached. I shim behind the arm plates because it is critical that the arms stay parallel to the ladder and that the pivot plates remain stationary. If they don't, the rivets that hold the arms will wear out from twisting and torquing as the stairway is used.

After I've screwed through the piano hinge and the arm plates, I shut the door and make sure there is an even reveal between the door and the jamb all the way around the door. When this is done I shim and screw off the rest of the wood frame, using #8, 3-in. wood screws.

Cutting stairs to length—Because ceiling heights vary, folding attic stairways come in different lengths, and with the exception of aluminum models, you must cut the bottom ladderlike section to length when installing the stairway. It is not difficult to figure out the cut length, but it is critical to the longevity of the stairway that the length be exact. A stairway that is cut too long will not extend to a straight line, and the ends of the ladderlike sections will not butt together (right photo, p. 34), putting undue stress on the hinges. And a stairway that is cut too short will stress the arm plates, the counterbalancing springs and the section hinges.

To cut the bottom ladder section to length, I make sure the arms are fully extended and fold the bottom section underneath the middle section. I rest my leg against the stairway to ensure that it is fully extended, and I take my tape measure and hold it along the top, or front, edge of the middle section (left photo, p. 34). By extending the end of the tape to the floor (while holding the upper part of the tape against the middle section), I get an exact measurement from the floor to the joint between the two lower sections. I repeat the procedure on the back edge of the stairway to get the length of the back of the cut. Then I remove the lower section, transcribe the measurements and draw a line between the two points on each leg.

After making my cuts and reattaching the bottom section of the stairway, all that's left to do is unscrew the temporary ledgers from the ceiling and run the precut trim around the frame. I've installed quite a few folding stairways, and

Measuring for trimming. With the bottom section folded under the middle section, the author puts his weight against the stairway to ensure it is fully extended. He measures along both the top and bottom edges of the stairway, transcribes his measurements on the bottom section, connects the dots and makes his cut. The trap door's pull cord can be seen hanging in the top of the photo.

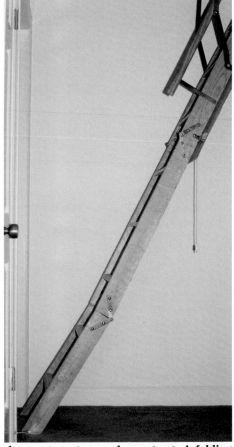

Accurate cuts are important. A folding stairway that is cut too long puts undue stress on the hinges because the ladderlike sections don't butt at their ends.

I can usually manage to do the whole job in about two hours.

Sliding stairways—Several companies make folding stairways, but Bessler Stairway Company also makes a sliding disappearing stairway. Unlike a folding stairway, where the sections are hinged and fold atop one another, the sliding stairway is one long section that slides on guide bars aided by spring-loaded cables mounted in enclosed drums. When the stairway is closed, the single-section stairway extends beyond the rough opening into the floor space above. This is an important consideration because some small attic spaces do not have enough room for the stairway's sliding section.

To access a sliding stairway, you simply pull the door down from the ceiling, similar to the way you'd pull down a folding stairway. Then you grab the single ladderlike section and slide the section toward you, lowering it to the floor. To close the stairway, you slide the single section back up into the opening. A unique cam-operated mechanism locks the ladderlike section in place while you push the door back to the ceiling. A series of spring-loaded, counterbalancing cables makes the door and the ladderlike section feel almost weightless.

The real benefit of sliding stairways is their angle of incline. Folding stairs typically have about a 64° angle of incline. That's pretty steep—more like a ladder than a staircase. Bessler's best slid-

ing stairways have a 53° angle of incline. Sliding stairways, unlike folding stairways, are designed so that the user can walk up into the attic while carrying a load (left photo, p. 32).

Sliding stairways are made of knot-free southern yellow pine, and there are four different models from which to choose. The smallest—the model 20—has a rough opening of 2 ft. by 4 ft. and has a suggested load capacity of 400 lb. This model has a stairway width of 17 1/16 in. The model 100 requires a rough opening of 2 ft. 6 in. by 5 ft. 6 in. and has a suggested load capacity of 800 lb. The width of the stairway is 18 7/8 in. Sliding stairways are measured from floor to floor, rather than from floor to ceiling like folding stairways, and the largest model 100 will service a floor-to-floor height of 12 ft. 10 in. Sliding stairways also have a full-length handrail.

The smallest sliding stairway, the model 20 with a maximum ceiling height of 7 ft. 10 in., costs around $225. The largest model, the model 100 with a maximum ceiling height of 12 ft. 10 in., costs around $700.

Installing sliding stairways—Sliding stairways do not come from the factory as assembled units; installation of these stairways is more for a journeyman carpenter because the finished four-piece jamb is not furnished and must be built on site. Stringers and treads need assembly, and the door and all hardware have to be installed on site.

I frame the rough opening 2 in. larger than the door opening. This allows me to use 3/4-in. stock for the jamb and still have 1/4 in. of shim space on each side to account for possible framing discrepancies. I rip the jamb stock to a width equal to the joist plus finished ceiling and attic flooring material.

It's possible to attach the finish trim to the jamb while it's still on the floor and then mount the whole unit into the rough opening using braces (called stiff legs or dead men) to hold the jamb to the ceiling while its being shimmed and nailed. But because I work alone, I screw ledgers to the ceiling the same way I do for folding stairs and then apply the trim later.

I nail the hinge side of the jamb to the rough framing and then hang the door with #10, 1-in. pan-head screws. Next I close the door to fine-tune the opening. After eyeballing the crack along the door edge, I move the jamb in and out to produce an even reveal down each side and then shim and nail the jamb.

Next I lay the stringers on sawhorses and thread the ladder rods with washers through the center holes of both stringers so that the stringers will stand on edge. Ladder rods are threaded rods that go under the wood treads, giving strength and support to the sections. I install all but the top three treads into the gains (or dadoes) in the stringers, screw the treads to the stringers, then tighten the nuts on the ladder rods. I always peen the ends of the ladder rods to keep the nuts

Right photo, this page: William T. Cox

Sources of supply

American Stairways, Inc.
110 Auction Ave.
Memphis, Tenn. 38105-1612
(901) 521-1100
American makes three models of folding disappearing stairways. The smallest has 1x4 treads and stringers and a rough opening of 22 in. by 4 ft. The largest has 1x6 treads, 1x5 stringers and a rough opening of 2 ft. 6 in. by 5 ft. Scissor hinges join the ladderlike sections. Optional accessories include an R-6 insulated door panel, bright orange rubberized painted treads and a fire-resistant door panel.

Bessler Stairway Co.
110 Auction Ave.
Memphis, Tenn. 38105-1612
(901) 522-9017
Bessler is a division of American Stairways, Inc. Bessler makes a folding stairway as well as a sliding stairway that has a one-piece stringer and slides on guide bars counterbalanced by spring-loaded cables. Bessler's folding stairway has high-quality section hinges that butt when the stairway is opened. Standard features include 1x6 treads and 1x5 stringers, as well as an R-6 insulated door and bright orange rubberized painted treads.

Hollywood Disappearing Attic Stair Co., Inc.
9525 White Rock Trail
Dallas, Texas 75238
(214) 348-7240
Hollywood makes the Wonder Action stairways that have two ladderlike sections that fold onto one another with an action similar to a parallel ruler.

Memphis Folding Stairs
P. O. Box 12305
Memphis, Tenn. 38182-0305
(800) 231-2349
Memphis Folding Stairs makes folding stairs that are very similar to the ones offered by American and Bessler. In fact, a person who worked for Memphis Folding Stairs now owns American Stairways. They also sell an aluminum folding stairway, as well as a heavy-duty wood model with 2x4 rails and 2x6 treads. Memphis sells a thermal airlock for its stairs that covers the stairway opening. It operates like a roll-top desk and has an R value of 5.

Precision Stair Corp.
P. O. Box 2159
Morristown, Tenn. 37816-2159
(800) 225-7814
Precision makes metal folding stairways and a fixed aluminum ship ladder with a 63° angle of incline. The company also makes an electrically operated commercial-grade sliding stair that has a switch at both the top and bottom of the stairway.

R. D. Werner Co., Inc.
93 Werner Road
Greenville, Pa. 16125-9499
(412) 588-8600
R. D. Werner is a large ladder manufacturer that also makes the Attic Master, which is its line of folding stairs. Of particular note is its aluminum stairway with adjustable feet and a load capacity of 300 lb.

Options include a wood push/pull rod that takes the place of a pull cord, self-adhesive antislip tread tape and a stairway door R-5.71 insulating kit.

Therma-Dome, Inc.
36 Commerce Circle
Durham, Conn. 06422
(800) 894-8589
Therma-Dome offers two insulating kits for attic stairs (R-10 and R-13.6) that consist of foil-covered urethane foam boards and touch-fastener tie-downs. These covers seal to the attic floor with a foam gasket. With their high R-values, payback will be quicker in colder climates. The covers cost between $65 and $80. Therma-Dome will fabricate covers for most stairways.

Trico Metal Manufacturing
266 Madison Ave.
Memphis, Tenn. 38103
(901) 527-5371
Trico manufactures three different grades of wooden folding stairways. —W. T. C.

---

from falling off. It's important to leave out the top three treads so that I can slide the ladderlike section onto the guide-frame bars at the top of the finished jamb.

When I install the guide frames and the two mounting brackets for the drums that contain the springs, I always predrill all of the holes with a ⁹⁄₆₄-in. bit. After 30 years, you would be amazed to see how the wood pulls away from where the screws were put in without predrilling. This causes a minute split to start, and when I repair sliding stairs that are 30 to 50 years old, the cracks have grown enough that I can stick a finger into them.

Installation of the mounting hardware is pretty straightforward. After putting the stringers onto the guide bars, I attach the cables. Caution: I wear gloves and am careful adjusting the cables' tension around the drums. If the cable slips, I could wind up like *The Old Man and the Sea,* with deep cuts in my hands and no fish dinner.

Another type of attic stairway—Hollywood Wonder Action attic stairways consist of two ladderlike sections mounted to a door in the ceiling. The stairways neither slide nor fold. The mechanical action of the Hollywood stairway is similar to a parallel ruler used by navigators and draftsmen. When the stairway is closed, the bottom section sits on the upper section. After pulling the door down from the ceiling, you lower the bottom section by pulling it toward you, just as you would a folding stair. But rather than

Hinged sections. Hollywood Stairways have ladderlike sections that are hinged like a parallel ruler. When opened fully the sections butt at their ends. On the right you can see the steel-tube handrails. When the stairway is opened fully, the handrails project 18 in. above the attic floor.

unfolding like an accordion, the bottom section pivots on four arms (two on each side of the stringer) and remains parallel to the upper section as you pull on it (photo above). When fully extended, the sections butt one another.

Hollywood has five models of stairways. The smallest, model 28-B, has a rough opening of 2 ft. 3 in. by 4 ft. 9½ in. and will accommodate ceiling heights from 8 ft. to 8 ft. 6 in. This model has 6-in. wide treads 17⅜ in. long. The largest model, the 45-A, has a rough opening of 2 ft. 9 in. by 6 ft.

3½ in., and it will accommodate ceiling heights from 11 ft. 1 in. to 12 ft. Model 45-A has 8-in. treads 23⅜ in. wide.

Hollywood stairways are sold as complete units, requiring only minimal assembly before installation. They are designed for residential and commercial use, and they are the heaviest of the disappearing stairways (115 lb. to 204 lb.) because they have solid ½-in. plywood doors and wide treads and stringers. The treads do not have ladder rods underneath them; rather, they are mortised into the stringers and fastened with wood screws.

Hollywood stairways have the same angle of incline as sliding stairs; they are just as easy to walk up or down while carrying things. Hollywood stairways have a unique tube-steel handrail that extends 18 in. above the attic floor. This gives the user more support than any other stairway. The stairways can be operated from the top, and this is especially useful for a second-story workspace when you want to pull the stairs up behind you (see *FHB* #43, pp. 70-71). Another nice feature of these stairways is that they come with mitered trim that's ready to be installed on the jamb. Hollywood stairways cost between $130 and $260, depending on the model. □

William T. Cox is a carpenter in Memphis, Tenn., who specializes in installing and repairing disappearing stairways. Photos by Jefferson Kolle except where noted.

Traditional Stairways Off the Shelf
About manufactured stair parts, and how they are made

by Kevin Ireton

On the banks of the Fox River in Oshkosh, Wis., stands an old stone building with a Victorian staircase just inside the front door. It's a heavy stair, done in dark oak, with a balustrade of spindles ascending on the left, complemented by a raised-panel wainscot on the right. The box newel that anchors the balustrade is also made of raised panels and is topped off by a square cap with a graceful curving cross section. The newel cap isn't nailed to the post; if you lift up on it, it comes off easily. Underneath you'll find a small scrap of sandpaper with writing on the back. In a faded penciled scrawl it says: "Built by R. W. Maurice Jan. 18, 1897." Such is the nature of stairbuilding and the pride it generates; people sign their work.

Stairbuilding is the pinnacle of the carpenter's trade. It combines the mathematical complexity of roof framing with the exacting standards of furniture-quality finish work. Traditionally, the staircase is the dominant architectural feature inside a house.

From the end of World War II through the early 1970s, during the heyday of the rambling ranch-style house, the practice of stairbuilding floundered, not only because the indiscriminate suburban sprawl encouraged one-story and split-level houses, but also because mass production and modern technology were streamlining construction techniques wherever possible—and one result was that wrought iron became the

A great deal of handwork goes into manufactured stair parts. Above, a worker at Morgan Products Ltd. uses a chisel to carve the inside of a volute where the shaper couldn't reach.

material of choice for many stair railings. But rising land costs have meant a return to multi-story houses. Also, the current popularity of old-house renovation has brought with it the need to restore and rebuild the stairwork of 18th and 19th-century craftsmen. All of this has fueled a resurgence of interest in traditional stairbuilding.

But the resurgence has been hamstrung by a lack of carpenters with an extensive knowledge of stairwork. Without an experienced stair-builder on his crew, a builder has two choices. He can either contract with a custom-stair outfit to build the stairway to specifications, or he can design the stairway around manufactured parts. Some other time, I'll write about custom stair-builders, but here I'll deal with how and where off-the-shelf stair parts are manufactured and sold. To find out about these exotic-looking

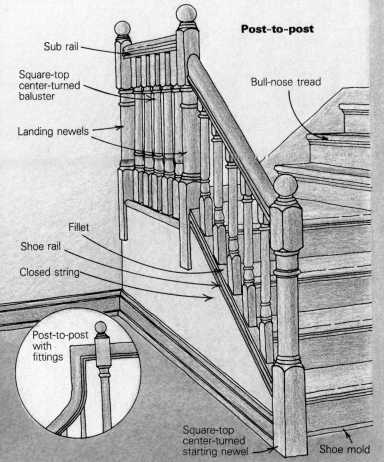

Post-to-post

Sub rail

Square-top center-turned baluster

Landing newels

Fillet

Shoe rail

Closed string

Post-to-post with fittings

Bull-nose tread

Square-top center-turned starting newel

Shoe mold

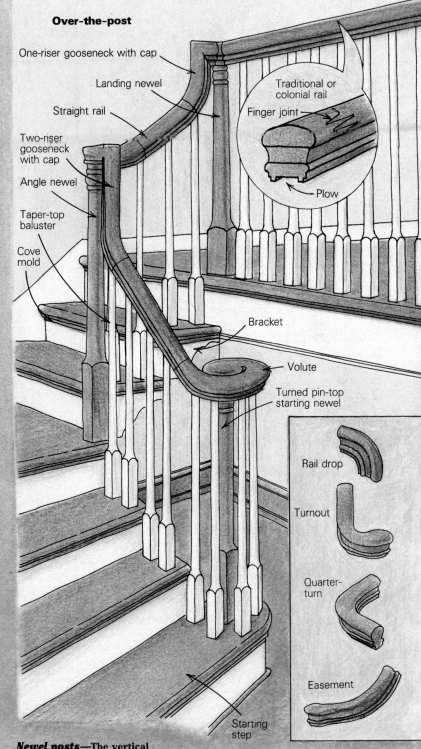

Over-the-post

One-riser gooseneck with cap

Landing newel

Straight rail

Two-riser gooseneck with cap

Angle newel

Taper-top baluster

Cove mold

Traditional or colonial rail

Finger joint

Plow

Bracket

Volute

Turned pin-top starting newel

Rail drop

Turnout

Quarter-turn

Easement

Starting step

Stairbuilder's lexicon

Balusters—Vertical members used to support the handrail and to fill the open area between the handrail and the tread or floor. *Taper-top* balusters (above right) are turnings whose shaft tapers, usually from 1¼ in. in dia. to ¾ in. at the top. *Square-top* balusters (above left) are used with plowed (grooved) handrail and can be uniformly square in cross section or *center-turned*, which means that the center section is a decorative turning.

Balustrade—The complete railing system, including newel posts, balusters and handrail. There are two basic types: in a *post-to-post* system (above left), the handrail is not continuous but is fitted between newel posts; an *over-the-post system* (above right) uses handrail fittings to create a continuous flow of handrail over the newel posts.

Brackets—Thin decorative pieces that are attached under the returned nosing of an open-string stair.

Closed string—A solid stringer that covers the ends of the treads and risers such that their profile cannot be seen (above left).

Fillet—A thin strip that fills the plowed (grooved) space between balusters in a handrail, sub-rail, or shoe rail.

Fittings—Sections of handrail used at the beginning and end of a balustrade or wherever the handrail changes height or direction. An *easement* is a fitting that curves in a vertical plane, used to change the angle of the handrail (shown at right is an easement with an integral newel cap). A *gooseneck* (above right) includes an easement and is used to change the handrail from the incline of the stair back to level, either at a landing or at the top of the stair.

A *quarter-turn* (right) is a level section of handrail used to make a right-angle turn. A *rail drop* is a curved fitting used as a decorative beginning or ending on a handrail (right). A *turnout* (right) is a starting fitting that curves in a level plane before curving vertically up the stair. A *volute* (right), or a wreath, is a starting fitting, similar to a turnout, that scrolls into a tight circle.

Handrail—The horizontal or inclined member that runs over the balusters and is supported by the newel posts. *Plowed* handrail (see detail above right) has a wide groove cut in its underside and is used with square-top balusters. *Wall rail*, usually simpler in design than handrail, is used where a stair runs along a wall and is attached to the wall with brackets.

Newel posts—The vertical members, larger than balusters, that support the balustrade. The *angle newel* (shown above) is the longest of the newels and is used at a landing where a stair changes directions and then continues to climb. A *center-turned newel* (above left) has *square ends* but a lathe turning in the center. A *box newel* is rectangular in cross section its entire length. A *landing newel* (shown above) is shorter than an angle newel and is used at a landing or at the top of a stair where the handrail changes from inclined to level. A *pin-top newel* has a dowel-like pin turned at its top, used to attach a fitting. A *starting newel* is simply the first newel of the balustrade.

Open string—A stringer that is cut out for the treads and risers such that their profile can be seen from the side (above right).

Shoe rail—A plowed rail that is used under square-end balusters when they don't sit directly on the treads (above left).

Starting step—The first tread and riser at the bottom of a stair, usually a step that is curved on one or both ends (above right).

Sub-rail—A thin rail attached to the underside of a handrail to make it more massive and allow for the use of square-top balusters (shown above left).

Drawings: Claudia Chapman

components, I visited some of the companies that make them.

Around the country, there are a handful of companies whose volume of sales qualifies them as major manufacturers of stair parts (see the sidebar on p. 41). I was able to visit three of them and tour their plants. I learned that these companies are highly competitive (at the distributor level), protective of their secrets and fiercely proud of their products. I also learned that, much to each other's chagrin, all three produce high-quality stair parts.

Stair-part basics—The balustrade of a staircase is the complete railing system, including newel posts, balusters and handrail. There are two basic types of rail systems—a post-to-post system uses square-topped or box newels with straight sections of handrail fitted between them, and an over-the-post system uses turned pin-top newels and various fittings—volutes, easements, goosenecks and quarter-turns—that allow a continuous flow of handrail (for definitions of the individual stair parts, see the previous page).

Most manufactured stair parts relate to the balustrade, but all the companies produce various other components as well. For instance, they all offer several styles and sizes of starting steps with curved risers. These are used with over-the-post balustrades that begin with a volute, and in some cases with a turnout. They serve as a structural and visual anchor for the balustrade, the shape of the starting tread reflecting the shape of the handrail.

Some companies make bull-nose oak treads for the rest of the stair, and these are available in various widths and lengths. You can also get nosings, cove and shoe mold, all to match the treads. Most of the companies offer decorative bandsawn brackets for use on the side of an open stringer stair. A few companies even make tread and riser caps for use along the outside edge of a stair that's to be carpeted. These give the illusion of a finished stair but save the cost of oak treads and risers that would otherwise be covered by the carpet.

Most of the industry, however, is devoted to producing newel posts, balusters, handrails and handrail fittings. The designs for many of these parts are very old, derived from the relatively delicate balustrades of colonial America. But some of the companies have recently come out with lines of stair parts that are more massive, including handrails as wide as 3½ in.

The three companies that I visited all have different product lines, though the most popular styles are available from all of them. They sell only through distributors, so if you contact them directly, they'll send you a catalog and the name of their distributor in your area.

Morgan—My first trip was to Morgan Products Ltd. in Oshkosh, Wis., where I met Sam Stecker, product manager for Morgan's line of stair parts. It was Stecker who showed me, with some ceremony, the signed note under the newel cap in Morgan's office building. After talking for an hour in his office, we walked outside into bright sunshine and crossed the street to the plant.

Morgan has been in business since 1855, when

it operated a planing mill on the same site that it occupies today. But now they're spread out over 27 acres and the plant is a labyrinthine conglomeration of buildings connected by walkways, conveyor belts, railroad tracks and even dirt paths. Some of the buildings are fairly new; others predate the 1897 note under the newel cap.

The stair parts that Morgan produces are either oak or birch, except for a small line of pre-assembled railings that are done in hemlock. They buy their lumber roughsawn, mostly from mills in the upper Midwest and Canada. Most of it is dried in their own kilns, a process that takes from 30 to 45 days depending on species, size and initial moisture content. Then the lumber enters the plant and is run through surface planers that give the operators their first glimpse of any defects. The next stages are crosscutting and ripping.

"Any manufacturer will tell you that a sawyer can make or break them," Stecker told me. The sawyers at Morgan use saws that rise up into the stock from beneath the tables. The planed and ripped pieces come to them on conveyor belts. The sawyers scan each piece for knots or other defects and gauge the distance between them. Next they slide the piece along under the saw guard, butting its end into one of a series of fixed stops on the other side of the saw. There are maybe a dozen of these stops, each for a specific stair part—baluster, newel or tread. The sawyer's job is to cut out all the knots, but at the same time, to get the longest possible piece.

From here the pieces for newels and fittings go on to gluing operations, where machines spread them with urea-base glue, clamp them and heat them at the same time to speed the curing process. The glued-up blanks are ready to be worked half an hour later. Pieces for turned newel posts and balusters are planed again to remove the excess glue and bring them down to final dimensions. Then these turning blanks are moved on carts to the automated back-knife lathes. Some of the baluster lathes are hand fed by an operator, but most are self feeding, holding up to 12 blanks at a time.

There are two basic types of balusters. Taper-top balusters are available with various amounts of decorative turning, but their distinguishing characteristic is that most of the shaft is a straight taper from about 1¼-in. dia. at the bottom to ¾-in. dia. near the top. Actually, the taper stops just below the top, and the last 3 in. is a uniform diameter. This is so that cutting the balusters to length doesn't affect the size of drill bit needed to install them in the handrail.

The other type of baluster is called square top or square end, and is used with a plowed handrail. The balusters are cut on the rake angle of the stair and toenailed into the underside of the handrail. The plowed areas between the balusters are filled in with thin strips of wood called fillets. The square-top balusters that Morgan offers are all center turned, which means that both ends are square but the center section has a decorative lathe-turned pattern cut into it.

When Stecker and I arrived at the station where newels are turned, the lathe operator was changing knives for a new run. He said the whole process took anywhere from two to four

hours. But once it was set up, he could turn out about 200 newels a day.

All the turnings go on to sanding machines where they get spun again, this time against rows of sanding belts of various widths (top left photo, facing page). These belts are backed by heavy broomstraws that press them against the baluster or newel.

Before being glued up into blanks, the stock for handrails is finger-jointed end to end to obtain the required length. I have worked with manufactured handrails and had always objected to the finger joints. Though strong and well done, they detract from the visual integrity of the handrail, especially when the adjoining pieces are markedly different colors.

I asked Stecker why they manufactured finger-jointed handrails. "We cut for yield," he told me. "It's a high-volume business. We manufacture miles and miles of handrail every week, and there just aren't enough clear lengths of lumber out there in the new-growth trees." They're also concerned about straightness and stability. Long pieces of solid handrail are more liable to twist or warp than glued-up rails.

Another advantage of finger joints and laminations is that Morgan can use up wood that might otherwise be wasted. This economizing is evident all over the factory. Everything is used for stair parts that possibly can be, most of what can't be used is chopped up and burned, either to run the kilns or to heat the plant.

Morgan offers four styles of handrail, with various other options possible using a sub-rail. However, their fittings are available in only one style. The most popular handrail—at Morgan and elsewhere—is called colonial by some companies and traditional by others (drawing, p. 37, inset at right). Morgan's traditional rail, which they call M-720, is stack-laminated with three finger-jointed layers ⅞ in. thick. Individual pieces in each layer are seldom shorter than 24 in., and the top layer of the handrail is limited to three or four finger joints, depending on its length.

After the handrail blanks are glued up, they're fed into huge molders that turn out smooth finished rails in one pass. No sanding is needed.

I asked Stecker how quality control was maintained. "We have 70 quality-control inspectors for stair parts alone. All our employees have the authority to reject a piece of wood that comes to their station. No questions asked." And sure enough, at each station there was a small pile of rejects from the operator. Some of these get chopped up, some are used by the salespeople in their training sessions. Still others turn up around the factory. Defective newels become the corner posts on the flatbed carts used to move stock from one station to another. In several places, I saw tables with balusters for legs. When I visited the product-development offices, I found rejected newel caps with holes drilled in them doing duty as pen and pencil holders.

The carpenter shop—In the course of the tour, Stecker and I had worked our way toward the carpenter shop. Straight handrail, balusters and turned newels are all pretty much the products of automation; the operators only feed and tend the machines. But all of the fittings—vo-

Morgan's taper-top balusters are sanded on the machine above. They are spun at high speed against rows of sanding belts at the back of the machine. The belts are made in various widths to conform to the shape of the balusters and are forced against the baluster by stiff broomstraws.

Handrail fittings look more complicated to make than they really are. They all begin as simple shapes cut out on the bandsaw. The Morgan operator in the photo above is making level quarter-turns for use around a stairwell opening.

At Visador/Coffman, the shape of the volute allows the operator to pass it all the way around the shaper, thereby avoiding a great deal of handwork. In the photo below, he is using a template that holds two volutes at a time, allowing him to shape the inside curve of one and the outside curve of the other. He will make the final pass on the shaper head to his left.

At Colonial, there is a greater juxtaposition of old and new methods than at other companies. Using a bolted down Jorgensen clamp for a bench vise, this worker pares the sides of a one-riser gooseneck with newel cap.

lutes, easements, goosenecks, turnouts—are done largely by hand.

At no point are all of Morgan's stair parts being produced at once. They do short runs of a given piece, based on orders received and on projected needs. Some days people work on four or five different types or styles of fittings.

The blanks for the fittings are carted into the carpenter shop. Their first stop is a hulking 36-in. bandsaw (top right photo, previous page), where the operator traces a pattern on one or two faces of the blank, depending on whether he's making a fitting that curves in one plane (like a quarter-turn) or in two planes (like some of the easements). He cuts them out quickly, both hands guiding the cut; then one hand repositions the piece as the other sweeps the scrap into a nearby cart. The cart of blanks diminished appreciably in the 15 minutes that I watched. And the bandsawn parts grew in neat geometrical stacks.

The carts full of bandsawn pieces are then rolled over to the shapers. Each of the operators here has two shapers going at once—the first removes most of the waste and the second makes the final cut. For some of the fittings, like the up easement with an integral newel cap, a heavy steel template is attached to the top of the bandsawn part and rides against a fixed pin centered over the shaper head. Other fittings are nestled into wooden jigs and ride on bearings mounted above and below the cutters. Some parts are worked freehand.

Most of the fittings are made in two pieces and are assembled at Morgan in one of two ways. The volutes come with an easement already attached, so that you have a level scroll of handrail that unwinds into a short curved piece (the easement), which makes the transition from

level to the incline of the stair. These two pieces are attached with flat metal fasteners called clamp nails. The clamp nails have longitudinal barbs running along their edges and are tapered slightly so that as they're driven into place they draw the two parts tightly together.

The gooseneck fittings, which create the transition of the handrail at the top of the stairs from inclined back to level, are made with a gooseneck piece and a level piece that are mitered together using wood splines and wedges.

Before I began my tour, I had not imagined that anywhere inside of the factory I would find people with hand tools working pieces clamped in a bench vise. But once joined, the fittings are dressed by hand, using chisels, rasps and sandpaper (photo, p. 36).

I asked Stecker why the profiles don't match exactly when you join a section of straight rail to a fitting. He pointed out that I had just seen the reason: the straight handrails are made by molding machines and the fittings are made on shapers. The cutters of these machines are changed all the time. Despite every attempt to keep the profiles the same, minor differences happen.

I spent two days at Morgan Products, Ltd. and developed a general understanding of how they make stair parts. I decided to spend one day each at the facilities of two other companies to find out if they were doing anything different. It turned out that the milling procedures that I saw at Morgan were more or less the same as at the other companies that I visited. But I did notice some interesting differences.

Colonial Stair Co.—The second company that I visited was the Colonial Stair and Woodwork Co. in Jeffersonville, Ohio, a tiny farming com-

munity in the central part of the state. Although Colonial began in Jeffersonville and is still headquartered there, the company now has two other plants, one 20 miles north in South Charleston and the other in Varney, W. Va. The Varney plant has all the manufacturing capabilities of the main plant and also has a sawmill. Colonial is the only company that I visited that cuts much of its own lumber from the log.

Forty years ago the building that Colonial occupies in Jeffersonville was a canning factory. Colonial started there because the building had a boiler, which they needed to run the kilns for drying the lumber. You can still see parts of the old canning factory, including the huge boiler that is stoked with sawdust by a man using a big coal shovel.

There is less automation at Colonial than at Morgan, fewer conveyor belts and more people handling the lumber. The plant is an old one-story building with a barn-like atmosphere. A lot of doors and windows were open on the hot June day that I was there. Some of the men had their shirts off. Two cats were making themselves at home, sleeping on workbenches, amidst all the dust and noise.

But it would be a mistake to assume from the small-town appearance that Colonial is anything but a major producer of hardwood stair parts. Paul Knapke, Colonial's vice-president and my tour guide for the day, pointed out that, calculated at the retail level, Colonial sold over $20 million worth of stair parts last year. And what's more important, when I examined one of Colonial's volutes, I couldn't tell the age of the building where it was made or that its pieces were crosscut by a 40-year-old Dewalt. I could see only that it was made with a great deal of care and skill.

Colonial also has some very modern equipment that belies the old-world atmosphere. They have two electronic gluing machines, for instance, that clamp the blanks and then cure the glue to near maximum strength in three to seven minutes. One of these machines holds up to ten 16-ft. long blanks for handrails.

At Colonial, the production of handrail fittings isn't confined to one part of the plant the way it is at Morgan. There is a small carpentry shop, with table saws, sanders and buffing machines. But the bandsaws and shapers are in other parts of the plant. The fittings are cut to length in the carpentry shop, assembled and for the most part hand-finished. I saw four or five people working in there, but they were all sanding or buffing when I visited, except for the man in the photo above left.

Colonial sells a videotape that they developed, called "Treasures in Hardwood." It's a two-part, 60-minute tape about stairbuilding and balustrade systems and sells for $54.95, including handling (the company's address is listed in the sidebar on the facing page). I wish the production quality of the tape were higher and that there was less time devoted to company history. But the information on the tape is useful, and although much of it is available elsewhere (all of the manufacturers furnish pamphlets on how to work with their stair parts), this videotape might serve as a good introduction for people who have never worked with manufactured stair parts before.

At the end of my tour, I asked Knapke if there was really a difference in quality between the products of the major manufacturers of stair parts. "I'd be lying if I said there was a big difference," he told me. "I'd like to think that maybe the quality of wood that we use is a little better because of the control we have over cutting and drying it."

Visador/Coffman—Without a doubt Visador/Coffman in Marion, Va., has the newest and largest plant of the companies I visited. Built in 1977, it has over 100,000 sq. ft. of floor space in the one-story building that houses the milling operation. Raw lumber goes in one end and finished stair parts come out the other. There are 2,000,000 bd. ft. of lumber air drying in their yard, and another 750,000 bd. ft. stored in a climate-controlled warehouse that reduces moisture content much faster than air drying. Then the lumber goes into their kilns. They move 35,000 bd. ft. through their factory every day.

When I arrived at Visador/Coffman, I spent some time talking with Bill Foster, the general manager, and then met John Gray, the resident engineer, who conducted my tour. I saw more technology and less handwork at Visador/Coffman than I saw at either of the other companies. But I noticed other differences too. For instance, their volutes don't draw into as tight a curve as Morgan's. I didn't like them as well, but the design allows Visador/Coffman to work the shaper all the way around the volute, eliminating any rasp and chisel work (bottom photo, p. 39).

Visador/Coffman has six molding machines that were all turning out handrail on the day I visited. These machines make so much noise that they're housed in little flakeboard cubicles. The blanks are fed into a small opening on one side, and the finished rail comes out an opening on the other. One of the older men who was receiving the finished rail as it came from the molder had his right hand loosely wrapped around the rail, letting the machine feed it through his fingers. "He's checking for defects," said Gray. "If there's the slightest variation, he'll feel it." Other workers I saw used a steel template to check the rails.

The most exciting thing I learned at Visador/Coffman is that they offer veneered handrail in several styles. This process, called profile veneering, has been done in Europe for many years but Visador/Coffman is the only U. S. manufacturer that does it. They buy the veneer in rolls, like toilet paper, and wrap it around finger-jointed handrails. The result is a length of rail that appears to have no finger joints. It's made here in three different styles, and the core is essentially a smaller version of their regular handrails, except that more finger joints are allowed since they won't show.

The chief criticism of this veneered handrail has been that when you attach it to a fitting and have to dress it down you are liable to sand through the veneer. But Visador/Coffman makes their veneered handrail slightly smaller than the fittings so that you can dress the fittings down and not have to sand the veneer. They also price the veneered handrail 15% cheaper than their regular handrail.

Another interesting item made by Visador/Coffman is bending rail, which you can use to make a handrail that conforms to the shape of a curved stair. Available in two styles, this rail is sliced vertically into layers about ⁵⁄₁₆ in. thick. To make a curved handrail, you use the stair itself as a mold, spreading the individual layers with glue and clamping them directly to the treads with the aid of L-shaped clamping brackets. Along with each bending rail, Visador/Coffman includes a 3-pp. illustrated installation guide.

Visador/Coffman also markets a couple of jigs designed by John Gray to help people cut and connect handrail fittings. One is a miter jig ($24.50) to help you cut fittings on your miter box. The other is a slotting jig ($120.77) that allows you to connect fittings to straight rail using clamp nails, instead of the traditional method, which is to use a rail bolt.

Designing and ordering—Though it may seem an obvious point, you should design your staircase, including the balustrade, before you build it. Some people assume they can find parts to accommodate whatever they design, and construction can get pretty far along before they go looking for some of the parts and fittings. Despite the wide selection of parts available, finding exactly what you need isn't always possible. Make the staircase a part of your house plans from the beginning. When you frame the stairwell, you begin to commit yourself to a specific design, and a lack of planning at this point can create problems that won't show up until later, when changes are especially frustrating and expensive.

Write to the companies listed above for their catalogs, and ask for any of their other literature (installation instructions, for instance) that would help you design your staircase. These catalogs contain lots of photographs and drawings that illustrate the design possibilities.

Once you've chosen a design, you'll have to make up a list of parts to order. This can be confusing since there are so many different parts and places to go wrong—left-hand or right-hand turnout, one-riser or two-riser gooseneck, angle newel or landing newel. You'll have to order your stair parts through the local lumberyard anyway, so contact them. They usually have a salesperson who regularly handles the stair orders and can help you determine the parts you need. If you're still confused, then call the manufacturer whose parts you want to order. All of the companies I visited said they get such calls regularly and are glad to help.

Order your stair parts as far in advance as possible. Although Colonial and Visador/Coffman both said they could ship orders in two weeks, that doesn't mean that your retailer can get them that fast. And even if he can, two weeks is still a long time if work has to stop because you waited until the last minute and then discovered that you had the wrong part.

Cost and compromise—The major manufacturers of stair parts sell only through distributors, who usually mark up the prices before passing them on to the retailers. And retailers of course

mark them up again before selling to you. I called Bill Foster at Visador/Coffman to try to get some straight answers about retail costs. He agreed that there are a lot of variables that make prices hard to discuss. But he doesn't know of any major manufacturers whose prices are out of line; he thinks they're all pretty competitive with each other.

Also, Foster gave me a list of suggested retail prices that Visador/Coffman furnishes to their distributors. Using that list, I added up the cost of parts for a traditional-style balustrade on a typical staircase with a single landing. I came up with $740 (no tax included) for all the newels, balusters, handrail and fittings.

The decision to use manufactured parts for a staircase instead of having them custom made is nearly always a matter of cost and convenience, just as it is with kitchen cabinets or any other architectural millwork that goes into a home. Manufactured parts are nearly always cheaper and more readily available.

While the decision to use them may be a compromise, a traditional staircase built with manufactured parts certainly doesn't look like a compromise. And given the challenges—designing it, ordering the parts and then building the stair—you won't feel you took the easy way out. You will feel the pride and satisfaction that increase directly with the challenges of any job. And maybe, as you set the last trim board in place, you'll feel the desire to sign your name on a scrap of sandpaper and slip it underneath. □

Installing Manufactured Stair Parts

Laying the handrail on the stairs is the first step

by Sebastian Eggert

In every cowboy movie there's a barroom brawl where someone crashes through the railing at the top of the stairs. If this could happen as easily as it looks, then our building codes would have outlawed wooden railings long ago. In reality, a well designed and built balustrade is not only beautiful, but it's also strong.

Like most builders today, I use stock parts for much of my stairwork. I buy the treads, tread-return nosing, newel posts, balusters and handrail over the counter (see pp. 36-41). All that's left is the challenge of installing them.

There are two types of balustrades that you can build. A post-to-post system has sections of handrail, either straight or with fittings, that fit between newel posts. This system is easier to install than the other. But for all the kids who enjoy sliding down the banister, it'll never do.

An over-the-post system uses straight rail joined to fittings, and flows gracefully over the tops of the newel posts (photo below left). This is the type of balustrade I usually build, and it's the one I'll discuss here.

The rough stair—Whenever I can, I frame the rough stair myself, rather than installing finished treads and risers over someone else's framing. This way I'm sure that the stairs will meet code requirements, and I avoid problems with carelessly installed carriages. At the same time, I can add blocking—for newel posts and wall-rail brackets—that I'll need later. (For more on stair construction, see pp. 8-14.)

After roughing in the stairs, I pull off the job until most of the other work is done. I like to get back just before the carpeting goes down. Since the stairs are often the only way to move between floors, I work the second shift if possible, to avoid interruptions by other tradespeople working on the house.

Lay it out in place—Some installation instructions included with stair parts tell you just enough to get you into trouble. In particular, the tables that list the heights of newel posts can be misleading. They work fine if your staircase is built just like the diagrams—same rise/run, same handrail heights, and most important, the same newel-post locations. But every installation is different, and most have their quirks.

The best way to build a balustrade is to lay out the handrail, fittings and all, in its exact location, right on top of the stair treads, the same way you lay out the plates of a stud wall (photo below right). Laid on the stairs, the handrail should touch the nose of each tread. The newel-cap fittings should sit directly above the centers of the newel-post locations.

Be aware that there is a procedural catch in the assembly process. The handrails need to be laid out over the locations of the newel posts,

When laying out a balustrade, measuring the components in place works best. Clamp the handrail and fittings in place right on top of the treads to determine where and at what angle to cut them. The finished installation shows a starting newel with a turnout and a two-riser gooseneck that had to be extended because of the location of the landing newel.

and those locations cannot be found easily without the treads in place. But the treads cannot be permanently installed until the risers and mitered stringers are on, and in some cases they can't go on until the newel posts are in. However, if the newel posts are in, you can't lay out the handrail on the treads. To cope with this confusion, I don't install anything permanently until the handrails are assembled.

Newel-post layout—Newel posts are the foundation of any handrail, and their placement is critical. Always plan for a newel post at the top and bottom of a run of stairs, and any place there is a landing or change of direction. Runs of over 10 ft. should have an intermediate newel post to strengthen them.

Except for starting newels with volutes and turnouts, which stand off to the side, newel posts should be laid out on the centerline of the balusters and handrail. Starting newels are usually notched around the corner of the first step. Newel posts at a landing or at the top of the stairs sit on the intersection of baluster centerlines or where dictated by the configuration of the fitting being used.

Turnouts and volutes sit on newel posts that are fastened to starting steps, which extend out from the staircase edge in a semicircle, with the newel post at the center of the circle. The manufacturers supply paper templates with these fittings to show the newel-post and baluster locations on the starting step. With all the newel-post locations marked, you can begin assembling the handrail.

Cutting the fittings—Handrail fittings are used to change the direction or slope of the handrail. Those used to change the slope include a curved piece called an easing or easement that has to be cut for the particular angle (the rake) of the staircase where it's being used.

I use a pitch block to mark the cuts on the fittings. A pitch block is a triangular piece, usually of wood, whose sides represent the rise, run and rake angle of the stair. One of the pieces cut out of the rough stair carriage will serve the purpose, or you can make a new pitch block from a scrap of 2x stock with the same dimensions as the rise and run of the stairway. If you're working on a stair with a landing and you didn't frame it yourself, check that the two flights of stairs have exactly the same rise and run. If they don't, make a separate pitch block for each flight to ensure accuracy.

At the bottom of the stairs, the handrail begins with a starting easing, turnout or volute. I set this fitting on a flat surface and snug the pitch block (run side down) to the underside where the easing turns up. Then I mark the point (tangent) where the hypotenuse of the pitch block touches the curve of the fitting (top photo at right). Then I turn the pitch block over on its short leg (rise side down) and scribe a line on the fitting (along the rake side of the pitch block) that passes through the first mark I made (photo at right, second from top). This gives the angle at which to cut the fitting.

It's hard to hold the fittings securely in a miter box, and at the proper angle to get an accurate

cut. Sometimes it helps to use the pitch block on the miter-box table to hold the fitting at the right angle. Fittings that have an integral newel cap can't be laid squarely against the fence, so for them I screw a piece of plywood to the bottom that acts as a jig (photo bottom right). With some fittings, like up easings, it's easier just to hold them by hand and hope for the best.

I use a Teflon-coated 80-tooth carbide cross-cut blade to make mirror-smooth cuts. Cut just shy of the line, and if the angle looks good, go for the final slice. Take too much and you've got a $50 piece of kindling.

To test the cut, clamp to the stairs a section of straight rail that has a square cut on one end, and while holding the newly cut fitting against the square cut, check the underside of the fitting with a torpedo level. If it reads between the lines, you're all right. Don't try to use your level on the top of the fitting because the millwork is rounded and not consistent enough to give a true reading.

Rail bolts—The next step is to join the fitting to a section of straight rail long enough to reach the next fitting location. I use rail bolts for the connections. These are 3½-in. double-ended bolts, with a machine-screw thread on one end and a lag-screw thread on the other. To locate the holes for the rail bolt, cut a wafer-thin piece of handrail (about ³⁄₁₆ in.) to use as a template. Drill a small hole through the template on the vertical centerline ¹⁵⁄₁₆ in. from the bottom of the handrail. Match up the template with the adjacent ends of the rail and fitting, and mark the hole locations with your pencil.

Drill a ¼-in. hole 2 in. deep in the end of the fitting and turn the lag-screw end of the rail bolt into it. To do this, I spin two hex nuts on the machine-screw end and lock them against each other, then use a wrench to turn the bolt into the fitting (photo at right, third from top). You can also just clamp the rail bolt with a pair of vise grips and turn it that way, but be careful not to damage the machine-screw threads.

Next, drill a ⅜-in. dia. hole at least 1 in. deep in the end of the straight rail. Then mark the bottom of the handrail 1⅜ in. from the end and on the centerline, and drill a 1-in. dia. hole 1½ in. deep. Be careful not to drill too deep or the point of the bit will come out the top of the handrail—another expensive mistake.

This last hole is the cavity where the nut is turned on the rail bolt's machine threads to pull the fitting and the straight rail together. After final assembly, you cover the hole with a wooden plug supplied with the rail bolt. I square up a portion of the hole facing the joint with a ½-in. chisel to provide a flat surface for the washer and nut to bear against.

Check the alignment of the fitting to the rail. If it's off, ream the ⅜-in. hole just enough to line the two up. Don't expect a perfect match. The cross sections of straight rail and fittings are always slightly different, so try to line up the bottoms and the side profiles as much as possible.

Most companies supply the rail bolts with star-shaped nuts that you turn with a hammer and a nail set. But the tapping upsets the alignment and levering mars the sides of the hole, so

These photos show a gooseneck fitting with a newel cap, quarter turn and an up easing. In the top photo the pitch block is being used (run side down) to locate where to cut the up easing. This procedure is the same as for any starting fitting, since at this point the up easing is the start of a new flight of stairs. In the next photo, the pitch block has been flipped over (rise side down) to mark the angle at which to cut the up easing. The photo above shows the fitting after it has been cut and with a rail bolt attached. The two hex nuts on the rail bolt were locked against each other and used to turn the rail bolt into the up easing.

Wherever the shape of the fitting makes it difficult to maneuver in the miter box, the author screws a piece of plywood to the bottom of the fitting as a jig. Here he's cutting a turnout easing with newel cap.

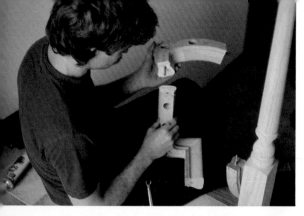

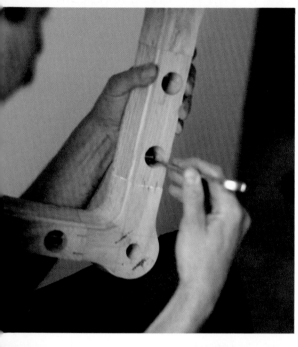

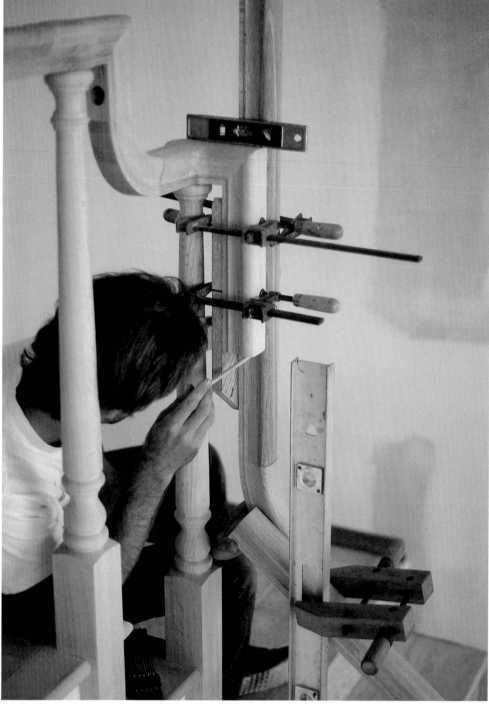

Ready for final assembly, the joining faces are coated with glue, the rail bolt inserted into the adjacent fitting (top left), and the nut and washer started, then tightened, through the large hole on the bottom of the fitting (middle left). Because they make for a smoother assembly, the author prefers hex nuts and a box wrench to the star-shaped nuts supplied with the rail bolts. The match between straight handrail and fittings is never perfect—some fairing is always needed. At left, a 2-in. dia. sanding disc attached to an electric drill is used to smooth the joint. In the photo above, the newel posts have been installed temporarily in order to calculate the cuts for attaching the easing between the gooseneck and straight rail. A section of straight rail has been attached to the easing and lined up with the vertical leg of the gooseneck to help position the easing correctly.

instead I use hex nuts and a twelve-point box wrench, as in the top two photos above left. This pulls them together carefully and firmly. But I don't glue them yet.

Goosenecks—Goosenecks form the transition from the incline of the stair back to level, either at a landing or at the top of the stairs. They come in two sizes, one-riser or two-riser, the difference being the length of their vertical leg. The size you need usually depends on whether the handrail simply levels off above the gooseneck or turns and continues to rise.

Sometimes two-riser goosenecks come disassembled so you can cut the vertical leg to whatever length you want. Single-rise goose-

necks never seem to be long enough, so I always use a two-riser gooseneck, just to be safe.

Instead of landings, some stairs have winders—treads that are wider at one end than at the other, used to effect a turn in a stair while continuing to climb. Winders are awkward and dangerous, so I try to discourage clients from using them. But if you do build them, be aware that they usually involve three rises at one point. You may have to add more straight rail to the vertical leg of the gooseneck to get enough height (photos, p. 42).

The length of the vertical leg of the gooseneck depends on the position and height of the newel post. The tables supplied with the gooseneck fittings tell at what length to cut the verti-

cal leg, but again, that measurement is correct only if the newel-post locations and handrail heights are the same as in the diagrams, so check them carefully.

If the installation lays out the way the instructions indicate, the leg can be cut to the assigned length, the up easing joined to the leg with a rail bolt, and the curve where the up easing joins the straight rail cut off using the pitch block.

Lay the assembled gooseneck on the bench top with the back of the leg flat on the surface, set the pitch block on its rise edge and slide it against the underside of the up easing to mark the intersection point. Then turn the pitch block over on the run side. Scribe the angle through the first mark on the up easing, and cut it there;

then attach the easing (without glue) to the gooseneck with a rail bolt.

You have to lay out everything in place again to determine where to cut the straight rail and attach the gooseneck fitting. Position the starting fitting over the newel-post location at the bottom of the stairs, with the straight rail attached and clamped in place on the treads.

Next you have to calculate the correct height to block up the gooseneck. If the height of the handrail coming up the stairs is to be 30 in., then the underside of the handrail will be 27¼ in. from the tread nosing (drawing, below). And if the height of the level handrail along the landing is to be 36 in., then the top of the newel post will be at 33½ in. The difference between these two measurements will be the distance from the underside of the gooseneck cap to the finished floor (33½ − 27¼ = 6¼).

Temporarily block up the gooseneck fitting this distance above the newel-post location. Now mark the point of intersection between the up easing you just cut off and the straight handrail coming up the stairs. Cut the straight rail, drill and attach the rail bolt, and assemble the gooseneck to the railing without glue. Now you can check all the fittings for alignment.

When the layout isn't exactly like the manufacturer's diagram, I stick with my empirical methods and lay everything on the stairs again, just as above, except that the easing isn't attached to anything. So now the straight rail is coming up the stairs and the vertical leg of the gooseneck is standing plumb at the top. The easing is simply an arc of a circle that's tangent to both of these rails.

Sometimes I can just hold the easing in place against the sides of the other pieces and calculate where to make the cuts by eye. But most of the time I attach a short section of straight rail temporarily to the easing and line this up with either of the other rails to find where to make the cuts (photo facing page, top right).

Newel-post installation—Once the handrail has been assembled, the newel posts can be cut to length and installed. With the assembled handrail lying on the stairs,

measure the distance between the bottom of the fitting and the stair tread below it. This measurement plus the height to the bottom of the handrail will be the height of the newel posts. Remember to add any additional length needed below the tread or landing to install the post.

The bases of the newel posts almost always have to be notched around the first step, into the corner at a landing, etc., so the alignment with the balusters is correct. Often the walls aren't plumb, so check them and lay out the notches accordingly.

I use ⁵⁄₁₆-in. hex-head lag screws long enough to reach the rough carriage or other framing members and pull the posts securely to the wall. With three lags per newel in different directions, they should never come loose. Plug the countersunk holes with the same 1-in. plugs used with the rail bolts and line up the grain so they disappear. At times I've had to pull the flat bottom of the newel post directly to the subfloor with a rail bolt, and count on the adjacent sections of rail and newel posts to steady it.

The newel posts manufactured for use with starting steps have a long pin or dowel turned on the bottom. Usually you can attach the newel to the step first, and then install both as a unit to the base of the stairs. Start by drilling a hole the size of the pin through the tread, then through the starting step's horizontal core. Secure the newel with a lag screw and washer through the bottom of the starting step into the newel pin.

Sometimes I fasten the starting step and tread temporarily in position, check to make sure that the newel post is plumb, and glue it into place. When the glue has cured, I remove the newel post

and step in one piece to screw them together under the framework; then I return the starting step and newel-post assembly to its location for final installation.

I've done some jobs where the pin-bottom newel post wasn't long enough. I had to buy a regular square-bottom newel and cut a tenon on the bottom. At every step, double-check everything and proceed with caution, especially if you're in the land beyond the diagrams.

When all the newel posts are securely fastened, all mitered skirt boards, risers and treads can be installed. As you fasten the return nosing to the tread, try to avoid putting the screws or nails where the tread has to be drilled for the balusters. I seem to ruin a drill bit on every job by running into my own screws.

Handrail assembly—With the newels in place, check the fit of the handrail assembly. You may have to rasp the pins on top of the newels so the handrail fittings will slide onto them without pounding. Check that the posts aren't pushed out of plumb by a rail section that's too long, and that the fittings are sitting level on the newels.

When I'm satisfied that the handrail is accu-

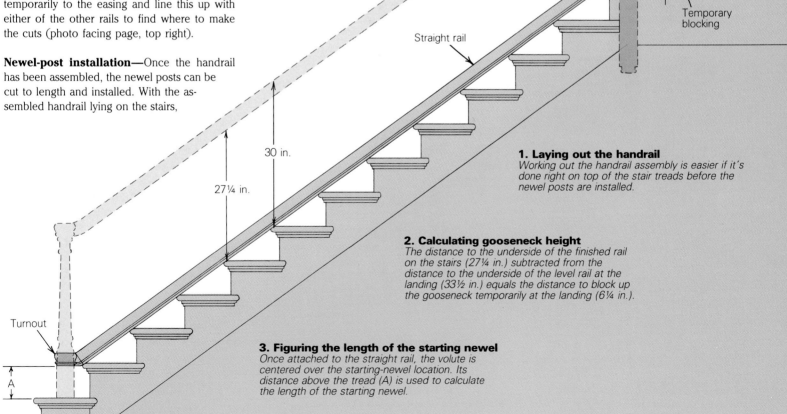

Gooseneck with cap

36 in.

33½ in.

6¼ in.

Temporary blocking

Straight rail

30 in.

27¼ in.

Turnout

A

Starting step

1. Laying out the handrail
Working out the handrail assembly is easier if it's done right on top of the stair treads before the newel posts are installed.

2. Calculating gooseneck height
The distance to the underside of the finished rail on the stairs (27¼ in.) subtracted from the distance to the underside of the level rail at the landing (33½ in.) equals the distance to block up the gooseneck temporarily at the landing (6¼ in.).

3. Figuring the length of the starting newel
Once attached to the straight rail, the volute is centered over the starting-newel location. Its distance above the tread (A) is used to calculate the length of the starting newel.

rately assembled, I pencil index marks across all the joints before I take it apart. These make it easier to line up the fittings when they're being glued. Where handrails level off, change direction and drop, the alignments are critical. A slight twist can throw off everything. If you're not sure that it will be a perfect fit when put in position, leave the joint dry and glue it together after all the balusters are in place.

If the handrail is fairly simple (one or two fittings), you can assemble the whole thing, plug the holes for the rail bolts, and fair and sand the joints before installing it permanently. If there are several changes in direction, you may have to assemble and install the handrail in sections. Assemble as much as possible beforehand, and plan the sequence of the final assembly to avoid problems tightening and sanding joints later. For instance, a joint at the vertical leg of a gooseneck and an up easing may be impossible to tighten in place, since the newel post will be in the way. Before the final assembly, I coat both sides of the joint with glue, then wipe off the excess with a damp sponge as soon as the parts are pulled together.

A heavy-tooth half-round file is useful for fairing the handrail joints, but I've also been using a small disc-sanding attachment on my drill to get into difficult places (photo, p. 44, bottom left). The discs I use are 80-grit and 2 in. in diameter. I've heard that a Dremel tool with a selection of small burrs also works well for this, though it might be tough to fair out the joints smoothly with such a small tool. I have used a belt sander, but usually just on the underside of the railing to smooth off plugs.

After fairing the joints, I finish the sanding by hand with 100-grit and then 120-grit aluminum-oxide production sandpaper until all of the scratches are gone. The stair-part manufacturers say that the newel posts and balusters are ready to finish, but don't believe it. Go over them all

just to make sure all the scratches are gone. I spin the balusters and newels on my lathe and sand them all with 120-grit paper for a really smooth finish. Don't use steel wool or sandpaper with a black grit on oak: it gets into the open grain and discolors it badly.

Baluster layout—Now that the newel posts are in place and the handrail is assembled, set the handrail up to mark the holes for the balusters. Traditionally, there are two balusters per tread, with the downstair face of the first baluster on each step in line with the face of the riser below it. The on-center spacing then should be half the run of the stair. I mark these locations on the treads, and plumb up to the handrail with a level or plumb bob, marking the underside for drilling the baluster holes. This layout can also be done when assembling the handrail on the treads, before the newel posts go in, simply by marking the baluster locations on the treads and then transferring the marks to the handrail with a combination square.

Some manufacturers suggest using a pitch block with a hole drilled in it as a jig to drill the holes in the handrail. But I just leave the handrail in place, line up the back of the drill motor with the mark on the tread below and drill carefully 1 in. into the handrail.

Sometimes a baluster layout will fall where a joint has to be made in the handrail. In this case, leave the baluster out until the joint is glued, plugged and allowed to cure. Then you can drill the hole for the top of the baluster. Cut the baluster just long enough to slide up into the top hole and drop into the hole in the tread. Glue and nail the baluster securely.

Always check the handrail for crowns or dips before cutting the balusters. I've had to pull crowns down by first toenailing a few key balusters and then pulling the rail down and toenailing the top of the baluster into the handrail. I

measure the overall length of each baluster, subtract a whisker or two for glue, and cut each baluster to fit in its place. Be sure that the bottom pin is short enough for the depth of the hole in the tread and cut it off if necessary. The balusters I use come with a tip diameter of 5/8 in. for the top 3 in., so they usually slide into the handrail without trouble.

Final installation—With the balusters all cut, the real fun is about to begin. I apply a good grade of construction adhesive like Max-Bond or Liquid Nails to all the holes in the handrail and treads, making certain that there's good contact all around. These glues seem to allow a bit of flexibility without compromising strength (unlike yellow glues, which tend to be brittle and don't fill voids as well). I also smear some glue on the pegs of the newel posts and balusters. Too much in the holes may prevent the balusters from sliding all the way in.

Set all the balusters quickly into the tread holes, spin them around once to make good glue contact, and then, starting at the bottom, begin setting the handrail in place. Trying to get all those balusters to line up and slide into place is like trying to get a roomful of children to sit still and pay attention. The ideal helper would be an octopus. It helps to slide each baluster up into its handrail hole, but without pulling its bottom pin out of the hole in the tread.

When all the balusters and newel-post pins are in their holes, gently but firmly tap the handrail down with a rubber mallet (photo below left). The balusters may jump a bit, so you have to check that their bases are tight to the treads and that they are lined up parallel to each other. Some pins are turned a bit off center, and those balusters may need to be rotated 90° or 180° to line up with the others. Check their alignment with a straightedge. Eyeball down the length of the rail for crowns and adjust with the mallet. Make sure the fittings are all the way down on the newel posts.

Sometimes I'll drift a long screw from the side of the fitting into the newel post to pull them together. Countersinking one through the top is easier, but if you do this be sure to find a plug that matches the wood grain around the hole and line it up carefully. Staircases are great rainy-day jungle gyms, so in houses with kids, I'll toenail all the balusters, top and bottom, just to be on the safe side.

After all the rails are on and the balusters are in, check for scratch marks from assembly and sanding. I use penetrating oil stains and clear finishes, applying several coats to bring up a smooth shine. Some parts like balusters can be prefinished, but wherever there is a nail hole to be filled or a plug to be sanded it can botch a nice finish, so it's better to finish the whole staircase after it's installed.

Take your time and do the best job you can. A beautifully crafted staircase is the centerpiece of a house, a delight to the eye and to the touch, and an example of the finest skill a carpenter has to offer. □

Lining up the balusters and worrying them into place for the final installation is a tricky job. It helps to have an extra pair of hands. Tapping with a rubber mallet ensures that the handrail and fittings are all the way down on the balusters and newels.

Sebastian Eggert builds staircases and mantels in Port Townsend, Wash.

Staircase Renovation

Loose, tilted flights signal trouble underfoot, but even major problems can be fixed with screws, blocks and braces

by Joseph Kitchel

There seems to be a great deal of reluctance among renovators to tackle any kind of stair rebuilding. I have seen many beautifully renovated homes with sagging staircases, loose balusters and makeshift railing supports. People who wouldn't think of leaving cracked plaster cracked or sagging floors unshored will put up with staircases so crooked they would make a sailor seasick, and squeaks so loud they wake the whole family at night.

Procrastination and lack of understanding of the mechanics of stair construction are at fault here. No one seems to know quite where or how to start, or who to call to do the job. To be sure, attacking an ailing stair takes courage, resolution and perseverance. It is one of the dirtiest and most disruptive renovating jobs.

The stairway is the spine of the multistoried row house. It links the sometimes minimal area of individual floors into what can be a spacious and accommodating floor plan. Because changing the stairway can seriously affect the physical flow of people and spoil the aesthetics of the whole interior design, it is far better to renovate than redesign or reposition.

Parts of a stair—Each step consists of a riser, the vertical portion that determines the height of each step, and the tread, the part on which you step. The dimensions of steps must be consistent within flights, though they may vary from one flight to the next. Variation within the flight will break the stair-user's physical rhythm and cause tripping. Tread widths may vary in the bottom two or three steps of the main flight for aesthetic purposes, and at the top of a flight, where pie-shaped steps are needed to negotiate a curve. Riser height, however, must not vary.

The newel post, usually decorated with paneling or carving, supports the railing at the bottom of the flight. Though it gives the impression of being heavy and sturdy, after years of being swung around by ebullient children, it has probably loosened enough to sway from side to side or lean out toward the hall. The vertical supports of the railing, aligned more or less behind the newel post and marching upward with each suc-

cessive step, are balusters or spindles, which are dovetailed into the treads.

Keeping the balusters from slipping out of their joints on the open side of the staircase are the noses, continuations of the molding that forms the front edge of the treads. These noses are removable and not part of the tread itself.

On each side of the stairs are the stringers, which hold the risers and treads. The stringer attached to the wall is usually routed or plowed out to receive the steps; this is called a housed stringer, and it produces a strong, dust-tight stairway. The outside stringer may be open or closed. Stringers may be simple, laminated with other pieces to form the curves of the stairs, or partially concealed by decorative filigree.

Were you to remove the plaster under the staircase, you might find two or three 4x4s or 4x6s running the length of the flight—one nailed to the inside of the outside stringer, one centered beneath the steps, and perhaps an-

other nailed to the inside of the stringer attached to the wall. These are the carriages, the members that add extra support. At the bottom, the carriages ideally rest on top of the stairway header, a joist that frames the stairwell end. They are sometimes attached to the inside face of the header. (This was the cause of failure in one flight I repaired. The weight of the stair forced out the toenails holding the carriage.) At the top, the carriages attach to the face of the upper-story header joist or, in the case of a curved flight, to angled braces running from notches in the wall to the upper floor joists.

Other elements of stair construction visible from beneath the flight are the wedges or shims driven between steps and stringers. Tapping these wedges tight or adding wedges made from shingles or building shims can do a great deal toward tightening up a staircase.

Diagnosis and dismantling—Problems with stairs fall into two categories. The simpler ones concern the railing, balusters, newel post and railing supports—the superstructure. Trouble here, though relatively easy to repair, can be symptomatic of more serious problems in the steps, carriages and stringers—the substructure.

Before you remove any plaster, explore the failings of the flight. If the steps are loose and tip toward one side, then the carriage or stringer on that side has weakened. The cause may be rotting, breaking, warping and splitting, or the carriage may have separated from the stringer.

Another symptom of major deterioration is a series of gaps or cracks along the ends of the treads or risers where they fit into the stringer. These rifts may occur on either side of the stairs, but are most often on the wall side, suggesting that a center or outside carriage has shifted downward, skewing the flight toward the center of the stairwell.

Large cracks in the plaster at the top or bottom of a flight are a good sign that the carriages have come loose from the headers. However, cracks generally parallel to the carriages or crisscrossing their length may indicate that vibration has caused the plaster to loosen and the keys to

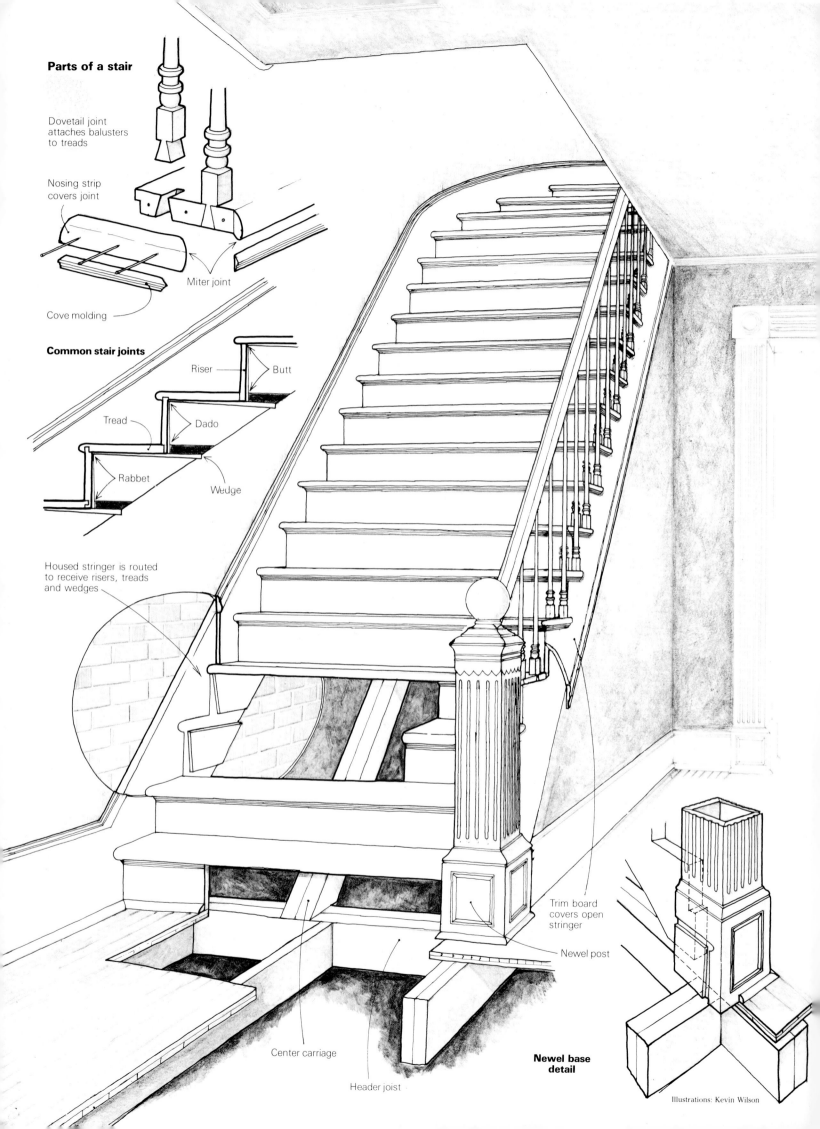

Parts of a stair

Dovetail joint attaches balusters to treads

Nosing strip covers joint

Cove molding

Miter joint

Common stair joints

Riser

Butt

Tread

Dado

Rabbet

Wedge

Housed stringer is routed to receive risers, treads and wedges

Trim board covers open stringer

Newel post

Center carriage

Header joist

Newel base detail

Illustrations: Kevin Wilson

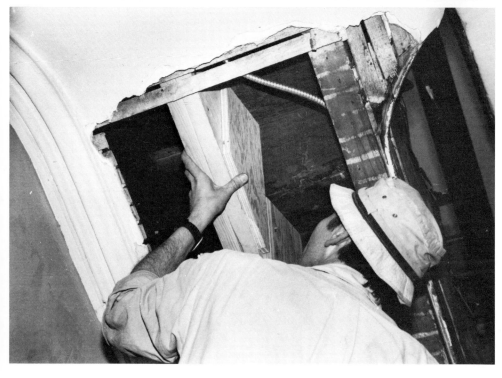

spindles, may have been purely cosmetic. Remove them so that all parts of the stairs may be properly aligned.

Substructure repair—You should now know the causes of your stair failure; tackle the bigger ones first. If a carriage is rotted out or cracked, tear it out and replace it. If it is intact but springy, it may be undersized; laminate a new beam or a steel reinforcing plate onto it. Solutions at this point must be as individual as the problems. But by far the most common failings are bolts and nails which have worked loose, allowing carrying elements to pull free from walls and header. In this case, mending plates made from $\frac{1}{8}$-in. steel work well.

If you must raise or remove carriages, first free them from each riser and tread; otherwise, the attached superstructure will loosen as the skewed members are jacked up. To remove treads and risers, pry them apart at joints or cut through the nail shanks if pulling is impractical.

For major repairs, jack the stringers or carriages into place in a manner that won't damage the superstructure and will keep it from damaging you. To give the jack or brace a level bearing surface, attach angle blocks cut to the slope of the stairs with clamps or screws.

break off, which is a much less serious repair.

If your stair has a decorative plaster molding running along its wall side, a continuation of the ceiling cornice, don't despair. This can be saved if it is not badly damaged and is still attached securely to the wall. To dismantle, carefully cut through the plaster and lath along the molding with a masonry blade in a circular saw, cutting parallel to the stair edge of the molding, leaving the molding intact. Plan this cut so that plaster or Sheetrock can be rejoined to this edge.

To further assess causes of stair failure, remove nearby plaster and lath. You may not have to remove all of the ceiling covering if the plaster isn't bad and if the problem is localized. Take time to clean up all the resulting plaster dirt. Debris allowed to accumulate on the flight below makes moving your stair platform difficult and dangerous, and will also worsen the

spread of dust throughout the rest of the house.

If necessary, remove noses and balusters, but label all parts first. Assign each step a number (I usually start at the bottom) and tape this number to the outside of the step and to its adjacent nose molding and spindle.

Using a screwdriver or chisel, gently pry the noses away from the stringer. They're usually finish-nailed in two or three places. You'll see the dovetail joint that connects the spindle to the tread. Tap the spindles out at the bottom and pull them down out of the railing.

Examine the joint between riser and tread to judge how to handle repair of squeaks or gaps. The joint may be a dado, a rabbet or simply butted and nailed. Past repairs, such as wedges driven into the gaps between riser and tread, fillers in the spaces above the treads along the wall stringers, or braces along the railing or

If the wall stringer has come loose from a masonry wall, reattach it by raising it to its proper position and nailing it with cut-steel masonry nails of sufficient length to go well into the wall. Wear goggles. A better method is drilling through the stringer and into the masonry with a carbide bit, and tapping in a lead sleeve; a lag bolt and washer expand the sleeve and tighten the stringer to the wall. If it's a frame wall, lag screws alone will hold stringers to studs.

If the outside stringer has twisted or warped, and is pulling the treads out of the wall stringer opposite, remove the treads and risers and force the stringer in toward the wall. Get the necessary leverage by temporarily bracing from the outside of the stringer toward the partition wall opposite. Screwing or bolting the stringer to the accompanying carriage will correct matters.

With the supporting members repositioned, now attend to the steps. Repair split treads and risers by removing them, lapping a piece of plywood over the back of the split, and gluing and clamping overnight. If you glue and screw the lapping piece you may forego clamping, and can replace the piece immediately. If the very edge of the nose is split, insert dowels from the edge to reattach it, being careful not to split the riser's dado or rabbet joint.

After you've corrected stringer and carriage problems and each tread and riser is back in place and renailed to the outside stringer, strengthen the steps by nailing step blocks to the center carriage. Why this wasn't done originally has always puzzled me. From a piece of ¾-in. plywood cut a step block to fit under each tread, and place it firmly against the back of each riser. Nail or screw the blocks to the side of the carriage, and then nail through the face of the tread and riser into the edge of the blocks. Trim the bottom edges of the blocks to conform to the angle of the carriage. I usually alternate the blocks on opposite sides of the carriage, but they all may be attached to the same side if the staircase is narrow and you can't get between the center carriage and the wall. Nailing blocks on both sides of the carriage is overkill, but add them wherever extra support is needed.

At this time a test run up and down the stairs will tell you where additional nailing and bracing are needed. For all face nailing in treads and risers, I use 6d or 8d finish-head, spiral flooring nails. For nailing where it doesn't show, I prefer 6d or 8d rosin or cement-coated box nails. I find 1½-in. or 2-in. Sheetrock screws driven with a variable-speed drill useful where hammer space is limited. Screws often add more strength than nails, because they pull things together and don't require pounding, which may disturb the alignment of nearby areas.

If risers, treads or stringers are to be refinished separately from the spindles, consider doing this now. Stripping, sanding and painting are easily done with the upper parts out of the way. Before finish is applied, set and fill all nails.

Superstructure repair—Newel-post problems are best dealt with after all other structural problems are solved. Although removing or loosening the newel post may be necessary to work on the carriages or stringers, it can usually be left in place to support the railing.

Newels are usually attached to the bottom step with a threaded rod, and to the railing with a hanger bolt. (Hanger bolts have wood-screw threads on one end and machine-screw threads on the other.) On the machine-screw thread of the hanger bolt is a star nut, which can be turned through the access hole with a screwdriver or needlenose pliers after the bolt is in place. (A plug fills the access hole later.) For extra strength screw into the railing from the inside of the newel post.

Straighten or tighten a shaky newel post by shimming under its bottom edges and renailing or screwing it to the floor. For greater support, try one of these repairs using a threaded rod. First, remove the newel post. Bolt a threaded rod to a bracket or wooden block so that the bolt fits flush with the bottom of the block. Screw the block and rod assembly to the floor, then slip the newel post over them and reattach it. To repair the post without removing it, attach two brackets under the tread of the first step as

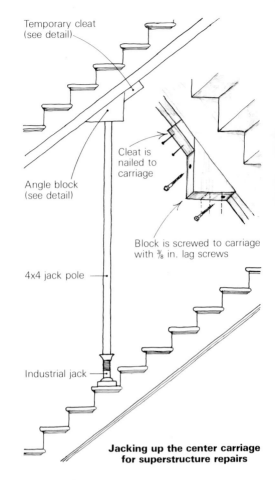

Temporary cleat (see detail)

Cleat is nailed to carriage

Angle block (see detail)

Block is screwed to carriage with ⅜ in. lag screws

4x4 jack pole

Industrial jack

Jacking up the center carriage for superstructure repairs

Building a Stair Platform

Before you renovate your staircase, you'll need to build a stair platform. Part of the reason stair repair is so difficult is that there is no place to stand, no way to get up to the job.

The exact dimensions of the platform depend upon your stairs. To make the platform, set a straightedge, level, on the fourth or fifth step from the bottom of your main staircase. The height of the platform is the distance from that step to the floor; its length is the distance from the back of that step to the front of the bottom step. Check these dimensions at different locations on various flights of your staircase. Sometimes rise and tread dimensions or angles of incline vary from flight to flight, but not usually. Then cut and attach legs and braces as shown in the drawing.

The platform should be wide enough to hold your stepladder comfortably, but narrow enough to allow passage on the stairs when it's in position. The platform will probably fit your neighbor's stairs, and you can use it when repainting your own stairwell; it is therefore a tool to retain after renovaton. You might consider making it collapsible for easier storage.

—J.M.K.

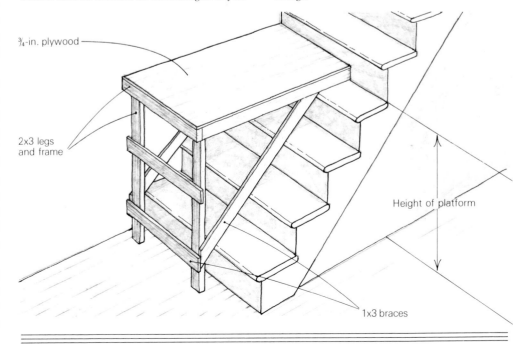

¾-in. plywood

2x3 legs and frame

Height of platform

1x3 braces

Newel-post attachment

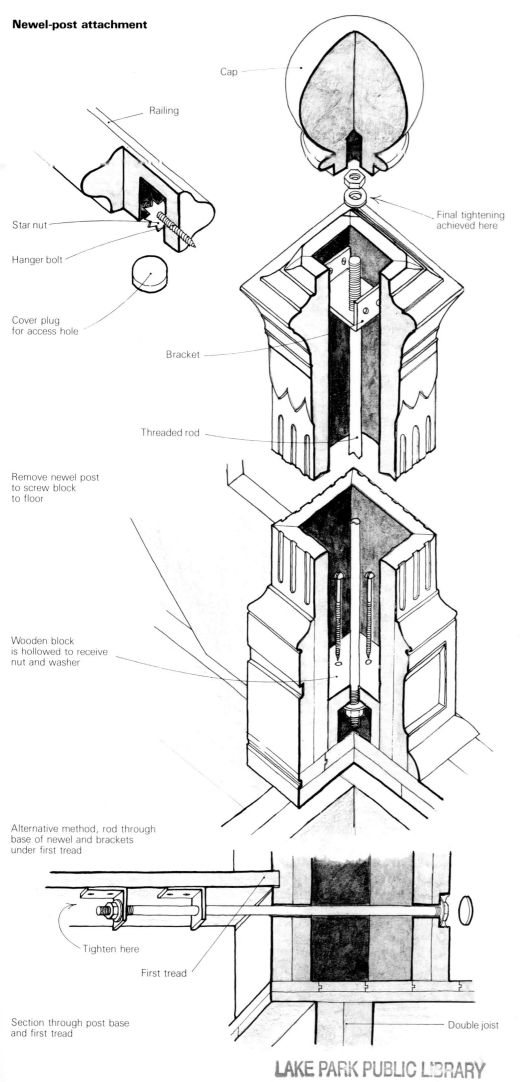

Cap

Railing

Star nut

Hanger bolt

Cover plug
for access hole

Bracket

Threaded rod

Final tightening
achieved here

Remove newel post
to screw block
to floor

Wooden block
is hollowed to receive
nut and washer

Alternative method, rod through
base of newel and brackets
under first tread

Tighten here

First tread

Section through post base
and first tread

Double joist

shown at left. Drill a hole through the base of the newel post to align with the holes in the brackets. Insert a threaded rod through the holes, bolt at both ends, and fill the access hole with a plug.

To replace the spindles, first coat both ends with white glue. Insert the top end of the spindle into the railing underside first, then slide the dovetailed end into its tread slot. Shim the joint wherever necessary, from underneath or from either edge, securing it with a finishing nail through the dovetail into the end of the tread. I find a rubber mallet useful when replacing spindles, because it will not mar the finished surface of the wood.

As you proceed from bottom to top, occasionally check the alignment of the railing and make adjustments by trimming or lengthening the spindles. Temporary braces hold the railing in place until all spindles have been installed and the glue is dry.

You can lengthen spindles (or adapt spindles from another stair) by adding a short piece of dowel. Drill into the top end of the spindle, and glue and nail the piece of dowel in. When the glue is dry, rasp and sand the dowel to the contour and taper of the spindle tip. Stain to complete the match.

When the final spindle has been inserted, the small bracket that originally connected the top of the rail should be reattached, or a new one made to fit. This prevents lateral movement of the rail and will hold the spindles in their correct positions as the glue dries.

Refinish the noses before you attach them. If you remove old nails by pulling them through from the back with a pair of nippers, you'll avoid the splitting that usually occurs when the nails are pounded back through and pulled from the face. Glue the noses and nail them twice along the side and once through the miter where nose meets tread molding.

Replacing the ceiling under the flight is the final step; use lath and plaster or Sheetrock. If your stairs curve, use short sections of wire lath to recreate the original curve.

Take care to keep all nailing surfaces in the same plane. This can be done with two straight-edges; one the width of the ceiling area, from the outside of the stringer to the wall, and the other as long as possible to run the length of the flight. Use building shims where necessary to keep furring strips in the correct plane. Determine your plaster line in relation to any plaster molding along the wall and to the bottom edge of trim pieces that adorn the outer carriage. I usually let the beaded or molded edge of this outer trim protrude below the plaster line. Existing pieces of plaster or plaster molding to be replaced may be drilled and secured with screws before touch-up spackling or painting. □

Joe Kitchel, 42, is a prop builder, cabinetmaker and renovator from Brooklyn, N.Y.

Installing Stair Skirtboards

Notching the skirtboard over the risers for a better-looking job

by Bob Syvanen

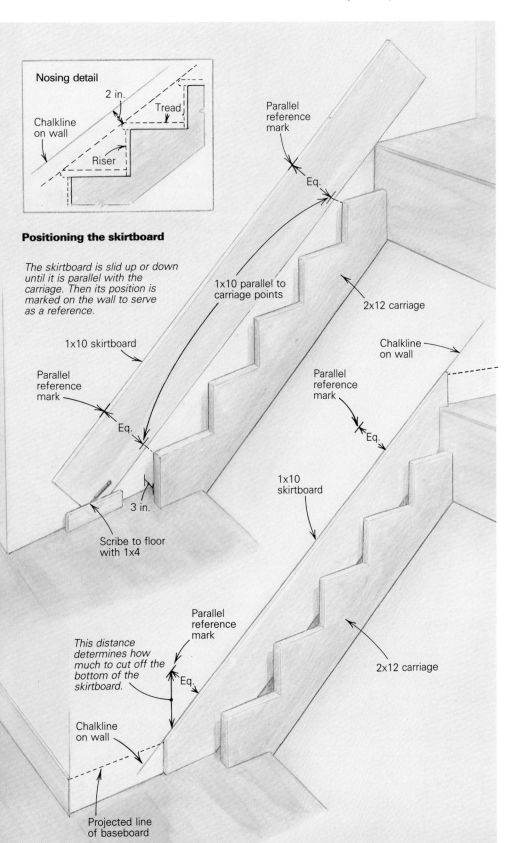

Nosing detail

2 in.

Chalkline on wall

Tread

Riser

Positioning the skirtboard

The skirtboard is slid up or down until it is parallel with the carriage. Then its position is marked on the wall to serve as a reference.

1x10 parallel to carriage points

1x10 skirtboard

Parallel reference mark

Eq.

3 in.

Scribe to floor with 1x4

Parallel reference mark

Eq.

2x12 carriage

Chalkline on wall

Parallel reference mark

Eq.

1x10 skirtboard

Parallel reference mark

This distance determines how much to cut off the bottom of the skirtboard.

Eq.

Chalkline on wall

2x12 carriage

Projected line of baseboard

I used to trim closed-stringer stairways by installing 1x10 skirtboards first and butting the treads and risers into them. Setting the first riser was simple enough—scribe, cut and nail in place. The difficulty started with the first tread. If it was the least bit long, it pushed the skirtboard, opening the joint below between the riser and the skirt. Installing the second riser then risked opening the joint at the first tread, and so on up the stairway.

A few years ago a young carpenter showed me a technique that involves notching the skirtboard for the treads and risers (it looks like an upside-down carriage). This method still requires scribe-fitting the treads, but eliminates scribe-fitting the risers because they slip behind the skirtboard. It produces a better-looking job quicker and with less aggravation.

Positioning the skirtboards—As an example, let's consider a straight-run stairway with walls on both sides. I install the outside carriages 3 inches in from the finished walls. This leaves access for work that must be done from the backside. Also, the underside of the stairway should be open (no drywall) for nailing and gluing access.

I lay a 1x10 skirtboard against a wall so that the bottom can be scribed to fit the floor (drawing left). If the carriage was against the wall, then I could rest the skirtboard on the carriage points. But I have to jockey the skirt a bit to make sure it's parallel to the carriage. With its bottom corner resting on the first floor and its upper edge resting on the second floor, I move the 1x10 up or down until the angle looks right. I check it by measuring the distance from the points on the carriage to the upper edge of the 1x10.

I tack the 1x10 in place and mark the wall along the upper edge of the board—one mark near the top and one near the bottom. These serve as parallel reference marks for determining the skirtboard's final position.

While the 1x10 is tacked in place I scribe the bottom end to fit the floor. The simplest way to do this is to place a short length of 1x4 flat against the bottom end of the 1x10. The top edge of this piece will be parallel to the floor. Marking along this line will provide the angle at which to cut the skirtboard. Later I'll calculate exactly how much I need to cut off.

After removing the 1x10, I return to the wall with the parallel reference marks and determine the top edge of where the skirt will eventually go. The skirtboard should sit as high as possible without exposing any wall where the riser and tread meet. The top edge of the skirtboard usually ends up about 2 in. beyond the finished tread nosings (detail drawing facing page). I snap a chalkline on the wall representing the upper edge of the skirtboard; I then use a level to draw the plumb cut at the top and bottom of the wall, where the baseboard will meet the skirtboard. I also draw a plumb line through one of the parallel reference marks and then measure along it between the reference mark and the chalkline. This distance determines how much I need to cut off the bottom of the skirtboard.

If I'm using 1x stock for baseboard, I'll extend the skirt as far as necessary to intersect it. Next I transfer the layout of the plumb cuts to the skirtboard. To do this I measure the height of the plumb line on the wall and use an adjustable bevel square for the angle. Then I find the point on the skirtboard where I can get the height I need at that angle.

I follow the same procedures for both skirtboards, make the necessary cuts and then tack them in place. At this point the skirtboard is between the wall and the carriage, and I'm ready to mark the cuts for treads and risers.

Marking and cutting— To locate the riser cuts, I use a length of riser stock, cut square at the ends, and press it up against the riser surface of the carriages and hard against the skirtboard (drawing above). I mark the location of each riser on the skirtboard with a sharp pencil. I do the same for the treads, still using the riser stock, marking horizontal tread lines. These tread lines, however, have to be re-marked ¾ in. lower to represent the bottom face of the treads (or the horizontal surfaces of the carriage). I re-mark them later.

You can use a circular saw to cut the skirtboards, but I prefer to cut them by hand. A sharp finish handsaw is best for the riser cuts. I back-cut them (cut them at a slight angle) for a good, tight joint at the face. The tread cut will be hidden so I use an 8-point crosscut saw here.

Running the risers— I like to rip all the risers to width before assembly. I keep the risers a hair narrower than called for and a bit short in length for a loose fit between the finished walls on each side of the stairway. I rip the treads to allow for a 1-in. nosing and crosscut them about ¾ in. longer than the dimension between the skirtboards. This allows me to slip them into position at a low enough angle to get a good scribe at one end while leaving enough stock to scribe and cut the other end.

When nailing the risers in place, I hold the top edge flush with the tread surface of the

carriages. After all the risers are in, I'll nail the skirtboards in place, but before doing that I locate the studs and blocking behind the drywall and mark the locations on the wall just above the chalkline that represents the top edge of the skirtboard. If there's a lack of solid blocking, I nail where I can and glue elsewhere with construction adhesive. Sticks wedged over to the opposite wall will secure the skirtboards firmly until the adhesive sets.

No matter how carefully the skirtboards are marked and cut, some joints will be open where they butt against the risers. This is easily fixed by driving shims between the riser and the carriage, and then nailing with 5d box nails through the back of the riser into the skirtboard. I do this nailing after all the risers are in.

Scribing the treads— Next the treads are scribed to the skirtboards on each end. Having allowed ¾ in. for scribing when I cut the treads to length, I set the scribes at ⅜ in. I put the tread in place with the end to be scribed down on the carriage and against the skirtboard (drawing right). The other end will ride high on the opposite skirtboard. The tread must be snug to the riser along the entire length. With the scribes riding against the skirtboard, I mark a line on the tread.

I back-cut the tread using a finish handsaw, but keep the cut square at the front where the nosing projects. Back-cutting makes it easier to correct the cut with a block plane.

After the first end is fitted, I carefully measure the distance between the skirtboards, using a folding rule with a sliding extension. Measuring for the length of either the front or back of the tread should be good enough, but I prefer to measure both as a doublecheck. With the tread in place, the scribes should be set to one of the marks and then should hit the other. If the scribe misses the second mark, it means the tread is tipped, and adjusting the tipping will ensure a good scribe line.   □

Bob Syvanen is a builder in Brewster, Massachusetts, and a consulting editor for Fine Homebuilding.

Marking and placing the skirtboard

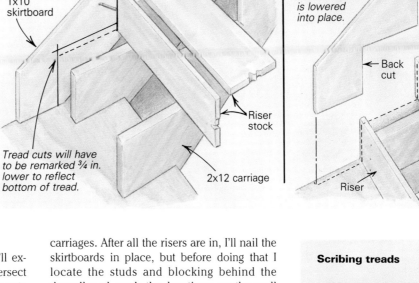

Hold riser stock against carriages to mark skirtboard for tread and riser cuts.

1x10 skirtboard

Tread cuts will have to be remarked ¾ in. lower to reflect bottom of tread.

Riser stock

2x12 carriage

After cutting the skirtboard and nailing the risers to the carriages, the skirtboard is lowered into place.

Skirtboard

Back cut

Riser

Scribing treads

1. With tread ripped to width and cut ¾ in. long, scribes are set at ⅜ in. to mark left side.

2. After scribing and cutting left side, measurements are taken to determine length of tread. Measure back and front with folding extension rule.

3. Right side is scribed to fit contours of skirtboard and marked for length at the same time.

Set scribes between skirtboard and measurement mark.

Left side already scribed.

Mark here from measurements.

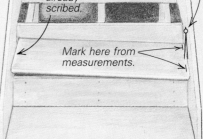

Drawings: Bob Goodfellow

Making a Bullnose Starting Step

You don't have to buy this classic detail; kerf-bending oak is easier than it looks

by Stephen Winchester

Most trim carpenters are comfortable hanging doors and casing windows, but when it comes to stairs, they call in a specialist. Stairs really aren't that difficult to construct and install if you take a bit of care, don't rush and make your joints as if the President himself were going to inspect your work.

On a recent job, I built a staircase with a double-bullnose starting step (photo right). The starting step is the first riser and tread on a stair. It's wider than the rest of the staircase and has curved ends, called bullnoses, that anchor the newel posts. In this article I'll explain how I made the starting step pictured here. Each staircase is different, so the dimensions I use here won't work for every stair, but the technique applies to any starting step.

Start with the tread—I have a small shop where I make treads, risers and moldings, but these components could also be made on site. I don't have a lathe, however, so I buy manufactured newels and balusters rather than make them. I also buy handrails and fittings. The stairs themselves, on the other hand, I usually frame on site, which is what I did on this project.

After installing the mitered finish stringers (the trim boards on the sides of the stairs) over the rough stringers and drywall, I measured the finished width of the staircase at the bottom step. On the starting step this is the dimension between the curved ends of the riser. The width of this stair is 42½ in. The bullnoses extend about 14 in. beyond the finished stringers, so the overall length of the starting tread blank is 70½ in. But I don't cut the tread to the exact length yet.

If I have a wide enough piece, I make my tread blank from solid stock. Otherwise, I glue it up from narrower stock and match the grain so that the tread doesn't look like a zebra. Manufactured tread blanks will work, but they're often glued up from narrower stock, and the grain doesn't always match. For this stair the run, or tread depth, is 10½ in. I added 1⅛ in. for nosing on the front and another 1⅛ in. for nosing on the back of the tread where the bullnose ends curve around. So I needed 12¾-in. wide stock. I use 1-in. thick stock for treads; ¾-in. stock is okay but looks a little flimsy.

To get 1⅛-in. nosing all around the starting tread, it's necessary to notch the back of the tread so that it fits over the second riser. The length of this notch equals the finished width of the stairs—42½ in.—and its depth equals the size

The first step. An open stair—where the finished stringers aren't boxed in by walls—makes it possible to have a double bullnose starting step. The riser is kerf-bent around forms that fit over the mitered stringers, and the tread is notched to fit in front of the second riser.

Photo this page: Rich Ziegner

of the nosing: 1⅛ in. When notched, the tread is 11⅝ in. deep where it hits the second riser and 12¾ in. deep only on the bullnose ends.

Cutting the bullnose—On this staircase the newel posts are capped by a spiraling section of handrail called a volute. The manufacturer of the stair parts (L. J. Smith, Inc., 35280 Scio-Bowerston Road, Bowerston, Ohio 44695; 614-269-2221) includes a paper template with the volute. The template helps you drill the starting step for the installation of the newel post and the balusters; I also used it to determine the radius of the bullnose ends on my starting step.

I cut out the corner of the template for the second-riser notch so that I could position the template on the starting tread (top photo, right). Using the outline of the volute as a guide, I added ⅞ in. to the radius of the volute, then used a compass to trace a half-circle on the template to shape the bullnose. Then I cut the template along the half-circle I drew and traced the bullnose on each end of the tread blank.

To cut the bullnose ends, I used my bandsaw, but a jigsaw will work on site. I clamped the tread to the bench and belt-sanded the saw marks. Then I used a ⅜-in. roundover bit in the router on the top and bottom of the tread to shape the nosing. After sanding, the tread was complete.

Making a bending form—Because I draw the shape of the riser on the tread, the tread becomes the pattern for the curved riser. I put the tread upside down on my workbench, set my compass to 1⅛ in. (the nosing dimension) and scribed around the edge of the tread. This mark created the shape of the riser that would support the tread. Because the riser stock is ¹³⁄₁₆ in., I scribed a second mark on the tread ¹³⁄₁₆ in. inside the first to create the inside dimension of the riser. I drew a line across the tread at each end of the second-riser notch to show where the mitered stringers butt against the tread.

Using the innermost mark, I measured the diameter and halved it to get the radius. Knowing the radius, I then made D-shaped bending forms for the curved riser out of kiln-dried framing lumber. Again, a bandsaw comes in handy for cutting the curved forms, but they could be cut with a jigsaw and belt-sanded smooth. I cut eight D-shaped blocks on the bandsaw and glued them together so that I had two stacks of four, each stack being 6 in. high. When the glue was dry, I trued up the curved edges on the bandsaw.

Next I set the glued-up blocks in place on the underside of the tread and screwed a plywood spacer to the blocks (middle photo, right). I turned the form over and screwed a second piece of plywood to the other side.

Kerfing the riser—After selecting a piece of stock from the pile of stair lumber, I made some sample kerfs—crosscuts that don't go completely through the wood—on the radial-arm saw (bottom left photo, right). When the kerfs are right, a riser will bend easily around a form, and the kerfs should just about close up on the inside of the bend. No heat or steam is necessary. My riser bend required that the ⅛-in. wide kerfs be cut

Forming the tread. A template is used as a pattern for cutting the bullnose ends on the tread. The cutout corner of the template aligns with the riser-skirt notch cut out of the tread.

Building the bending form. By scribing around the bullnose ends of the tread with a compass, the author determines the size and shape of the bending form. Each block is made from glued-up framing lumber; the blocks are placed on the tread and are held in place with a plywood spacer.

Kerfing with a radial-arm saw. Kerfs are crosscuts that don't cut completely through the stock. Here the kerfs are ⅜ in. o. c., and the face is ¹⁄₁₆ in. thick. Although the author used a radial-arm saw, a circular saw, a straightedge and elbow grease will yield the same results.

Perfectly kerfed. Kerfing begins before the form bends to eliminate tension in the riser stock; well-kerfed wood wraps around the form like a piece of paper. Here the kerfed stock is temporarily clamped so that the other bullnose end can be marked for kerfing.

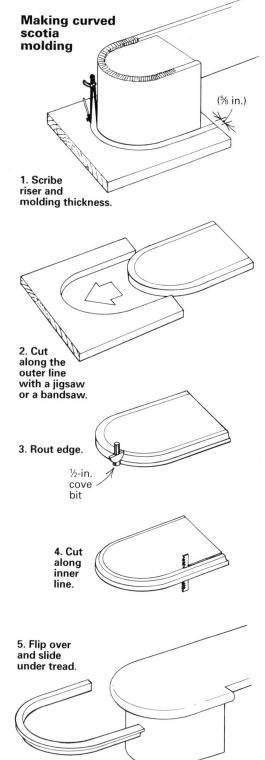

Making curved scotia molding

1. Scribe riser and molding thickness.

(⅝ in.)

2. Cut along the outer line with a jigsaw or a bandsaw.

3. Rout edge.

½-in. cove bit

4. Cut along inner line.

5. Flip over and slide under tread.

Hugging the curves. The riser is glued and clamped to the forms. Because the kerfed riser bends easily around the forms, the clamps are just snug enough to keep the stock in contact with the forms. Once the glue dries, the ends of the riser are trimmed to butt into the mitered stringers.

It fits. Careful measuring and cutting pay off when the completed riser slips over the mitered stringers. Construction adhesive and nails fasten the curved riser to the stringers; the tread is installed once the front of the riser is perfectly straight.

⅜ in. o. c. and ¾ in. deep. The uncut face of the riser ended up about ¹⁄₁₆ in. thick.

I tried a couple of different spacings before I got it right. I usually start with kerfs ½ in. o. c., then check to see if the wood bends smoothly around the form. You know you've got the proper spacing when the stock bends around the form easily. Remember, different species will not bend the same way. In other words, don't make trial cuts in pine when bending a piece of oak.

To measure the length of the piece I needed, I wrapped my tape measure around the forms and added about a foot for safety. This riser stock was almost 10 ft. long. I kerfed one end of the riser, beginning about 1½ in. before the start of the curve

so that the riser wouldn't kink as I bent it around the form. After clamping the kerfed end temporarily on the form (bottom right photo, p. 55), I marked the start of the second bend (plus 1½ in.) and cut those kerfs. A kerfed piece is pretty delicate and must be carefully handled.

Gluing and clamping—I set the form on the upside-down tread and held them together temporarily with a few spring clamps. The glued-up blocks are part of the installed riser, so I coated them heavily with yellow glue and drizzled some into the riser kerfs, too. Clamping one end with a pair of bar clamps, I quickly but carefully bent the riser around the form and clamped the other

end (top photo, this page). You don't need much pressure to hold the riser in place; that's why I didn't bother with clamping blocks.

I nailed the riser to the forms on what would be the back of the step. As soon as the riser was nailed to the forms, I took the riser off the tread. Otherwise, excess glue would have bonded the riser to the tread, and it wasn't time to put these two together yet. The next day, after the glue had set, I cranked my radial-arm saw way up to cut the extra length off the back of the riser. If you're doing it on site, use a circular saw and a square for this job. A 2° bevel on the cut ensures a tight fit to the finished stringers. The plywood spacers stay on until I am ready to install the riser.

Drawings: Bob Goodfellow

Making curved scotia molding—Scotia molding (or cove molding) is traditionally installed under the nosing of each tread on a stair. This molding hides the joint between tread and riser. To make the curved scotia for underneath the bullnose starting tread, I began by setting the riser upside down on a scrap of oak that was wide and long enough to make a U-shaped piece (drawing facing page). First I traced around the riser and then set my compass to ⅝ in. and traced around again. This second line is the outside edge of the molding. I marked a piece for each end of the riser, cut the outside curve on the bandsaw and then ran a router around each piece. The scotia profile is shaped with a ½-in. cove bit in the router. Then I went back to the bandsaw and cut the inside line. An identification letter or number helps me remember on which end of the riser each piece of molding fits.

Installing the step—After removing the plywood spacers from the riser, I set it in place to check that it fit against the finished stringers. I squeezed construction adhesive on the finished stringers and set the riser back in place (bottom photo, facing page). Then I drilled pilot holes and nailed the riser tight to the front of each rough stringer. I then held a straightedge to the face of the riser to make sure the riser was straight. (If it's not, a tap behind the low spot will make it straight. When the construction adhesive hardens, it fills the gap between the center stringer and the riser.) To pull the ends of the riser tight to the finished stringers, I ran a couple of screws through the inner faces of the rough stringers into the curved bending forms.

Next, I set the tread into beads of construction adhesive squeezed onto the tops of the rough stringers, the top edge of the curved riser and the bottom face of the second riser. I predrilled and nailed through the tread into the stringers with 10d finishing nails. Then I nailed the tread to the curved riser with 8d finishers. A couple of 6d box nails through the back of the second riser into the starting tread, along with plenty of construction adhesive, prevents a squeaky starting step. I don't use a nail gun to assemble the skirts, the risers and the treads because I could miss what I'm trying to nail into, or a nail could veer off and shoot through the face of the work.

The scotia molding goes on next. The two U-shaped pieces go first, followed by a straight piece (which I also make) under the front of the tread connecting the two U-shaped pieces. I use a small brad gun to shoot the molding on. It's easier than handling small nails.

To anchor the newel posts, I drilled a 1½-in. dia. hole in each end of the starting step, using the volute template to locate the holes. I drilled through the tread and the 2x bending forms. Then I cut the dowel tenons on the ends of the newel posts to length, coated them with construction adhesive and drove them into the holes until the base of the newels seated against the tread (see the sidebar on this page). □

Stephen Winchester is a carpenter and woodworker in Gilmanton, N. H. Photos by the author except where noted.

Installing a starting newel
by Bob Goodfellow

A starting newel provides most of the support for a handrail at the bottom of a stairway. A 1½-in. dowel tenon, turned on the end of the newel post, anchors the starting newel to a starting step. Wedging this dowel below the subfloor is the best way I know to install a starting newel so that it won't loosen up.

When you order your starting newel, get one with a dowel about 14 in. long so that it will extend clear through the starting step and the subfloor with length to spare. But before you bore a 1½-in. hole through the starting step and the subfloor, go downstairs and make sure there are no pipes, heating ducts or electrical wires in your path. Then begin work on the starting step. Establish the location of the 1½ in. hole on the starting tread by using the stair manufacturer's template. Pay careful attention to where you mark the hole because the starting newel's location directly influences the alignment of the handrail components.

If you drill carefully, the starting newel will sit plumb in the bore. If not, a little judicious reaming should rectify the situation. Make sure the tread is clean so that when you stick the dowel in the hole, the bottom of the newel is completely seated against the tread. Then go downstairs and mark the dowel where it penetrates the subfloor. Mark all the way around the dowel, then make a second mark to indicate the direction that the floor joists run.

The marks you made show you where to drill out a slot in the dowel for a wedge. The mark around the dowel locates the top of the slot, and the second mark shows you in which direction the slot should face (parallel to the joists). Make the slot by drilling a series of ⅜-in. holes. Drill the top hole a little above the subfloor mark on the dowel and then drill the others below until you've made a slot about 1½ in. long. Clean out the burrs of wood with a chisel and a rasp. Then mark the base of the newel to show which way the slot faces.

I make the wedge that anchors the newel from a piece of ⅜-in. oak stock. The wedge, which is about 6 in. long, has a taper on one side only; the other side I leave square. The back end of the

wedge should be about 1⅝ in. high, the front end about ⅞ in. high.

I also round over all the square edges; this way the wedge will have more contact with the slot in the dowel. Rounding over is done quickly with a block plane.

With the newel rotated according to the mark you made on its base, insert the dowel into the starting step. Then go downstairs with the wedge and your glue. Smear the wedge and the slot in the dowel with glue, then gently drive the wedge into the slot. It's a very good idea to have someone upstairs to yell, "OK!" when the newel is plumb and level. Don't bash the wedge in; you'll know when it's tight because it'll have a certain ring—dong, dong, ding—ah, that's it. Finally, drill a pilot hole and use a resin-coated nail to pin the wedge in place.

—Bob Goodfellow is a former stairbuilder and is now associate art director at Fine Homebuilding.

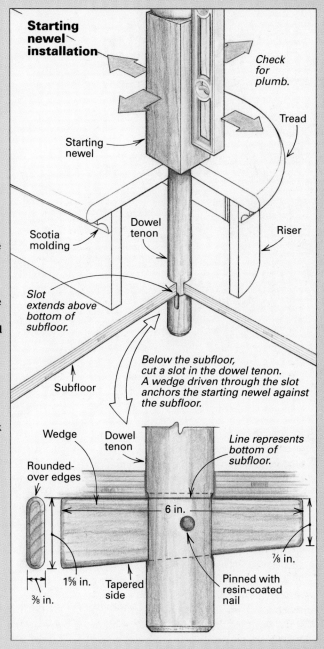

Starting newel installation

Starting newel

Scotia molding

Slot extends above bottom of subfloor.

Check for plumb.

Tread

Dowel tenon

Riser

Subfloor

Below the subfloor, cut a slot in the dowel tenon. A wedge driven through the slot anchors the starting newel against the subfloor.

Wedge

Dowel tenon

Rounded-over edges

Line represents bottom of subfloor.

6 in.

⅜ in.

1⅝ in.

Tapered side

⅞ in.

Pinned with resin-coated nail

A Staircase of Glass and Maple

Making the handrail was the hard part

by Scott McBride

John Raible practices architecture by trade and builds sculpture by avocation. That spells trouble for the person called upon to render his designs, which often lie somewhere on the borderline between the two disciplines. I built a house recently for one of Raible's clients. The second floor has an open floor plan under a tent-like hip roof. Two staircases, one from the entry foyer and one from the kitchen, lead up to a central stair hall that overlooks the public areas of the first floor. Daylight floods the hall from a number of skylights and cascades into the rooms below. To preserve the dramatic effect of openness and light, Raible called for an open-riser staircase in the foyer. And instead of newels and balusters, the continuous 3-in. dia. maple handrail was to be supported entirely by ¾-in. thick panels of tempered glass (photo facing page). Installing the glass was tricky enough, but creating the handrail was even more time-consuming. It called for a number of jigs and lots of patience.

An invisible balustrade—The plan was to lag bolt the glass to the stair stringers and to the framing that formed the stairwell, using wood blocks as washers to cushion the glass. Ledgers would bear the weight of the glass, which was considerable, so the bolts would only have to clamp the glass firmly in position.

The glass fabricator needed templates for the glass panels, complete with the location of bolt holes, so I decided to mount these templates in position first, build the handrail on top of them and then disassemble the whole affair. This would allow me to start work on the handrail almost immediately, and would let me easily reconcile any differences between field conditions and the drawings. The templates were made from ¾-in. plywood, equal to the thickness of the glass, and I reinforced them with 2-in. wide plywood ribs to ensure that they would stand straight when fastened. This was essential because I had to depend on the top edges of the templates to be an accurate guide for building the handrail.

To prepare the supporting surfaces for glass, I first screwed 1x2 ledgers to the lower edges of the stringers and the stairwell rim joists (drawing facing page). These ledgers would bear the weight of the glass. Next, I carefully shimmed the mounting surfaces at the location of each bolt. My goal was to align all the bearing points for each panel in a true vertical

plane. This had to be done with precision because any out-of-plumbness over the distance between upper and lower bolts would be amplified at the level of the handrail. Also, if the mounting points were out of alignment it would be easy to twist the glass to the breaking point when tightening the bolts. As it turned out, much of this painstaking effort was in vain. The glass panels warped somewhat during the tempering process, which involves baking the glass in a furnace, and this caused plenty of headaches in the final installation.

To keep the glass from cracking, I placed ¾-in. thick vertical blocks between the bolt heads and the glass. The blocks acted as soft washers, distributing the bolts' pressure over the glass. Each block was dadoed to create a recess for the bolt heads, leaving only ¼ in. of wood between the bolts' metal washer and the glass. The blocks also served as a nail base for the ¼-in. maple plywood cover boards which would conceal this fastening arrangement. Rabbeted horizontal trim pieces were nailed to the tops of the blocks to house the lengthwise edges of the cover boards.

Tooling up for the handrail—After bolting all the glass templates into position, I could determine the lengths of handrail I'd need and the number of elbows. Rather than attempt to develop specific falling easements for each change of pitch and direction in the handrail, I made a number of single elbows which, when used singly or in combination with others, would fit any of the conditions found in the two staircases. In principle it would be like making the handrail out of straight lengths of plumbing pipe and various fittings, except for the fact that the wood elbows could be readily cut to any angle between 0° and 90°. As long as these cuts were made through a true radial plane, the resulting section would mate accurately with another elbow or a straight run of rail. The ability to shorten and combine the elbows in this fashion would allow for a fairly graceful flow of the handrail—not so fine as a traditionally lofted handrail, perhaps, but not so crude as a pipe rail either.

To mill the 3-in. dia. handrail I needed some serious tooling. Although you can scout the catalogs and find a 1½-in. radius roundover bit for a router, the catalog description sometimes states that these bits are "not guar-

anteed." I'm not quite sure what that means, but with so much steel spinning at 22,000 rpm I'd just as soon not find out. Steve Schermerhorn, a sales representative from Chas. G. G. Schmidt & Co., Inc. (301 W. Grand Ave., Montvale, N. J. 07645), fixed me up with a hefty set of ball-bearing shaper collars and a pair of 1½-in. radius roundover knives milled from ⅜-in. thick high-speed steel (HSS). Corrugations on the upper edges of the knives lock them securely between the collars. Steve also gave me valuable advice on the design of the jigs I would need to make the elbows safely. "Build them heavy," he warned me, "and keep your hands far away from the cutter." I understood what he meant the first time I set the cutter spinning. It emitted an ominous-sounding "whoosh" and fanned a breeze that gave me a lump in my throat.

Making the straight run—Milling the straight runs of rail was simple enough. After dressing solid maple stock to 3 in. by 3 in., I plowed a 1-in. wide groove on the underside of the stock, using a straight cutter mounted in a table-saw molding cutterhead. This groove would receive the top edge of the glass panels (I allowed a little extra clearance that I could fill later with clear silicone caulk).

With the collar/knife setup mounted on the 1-in. dia. spindle of my shaper, I made the first, second and third roundover cuts, with the stock pressed firmly against the table and fence by a hold-down. To prevent the work from rolling as I completed the last roundover cut, I mounted a 1-in. wide guide strip to the outboard side of the fence (top photo, p. 61). This strip engaged the glass groove as the finished rail exited past the cutter.

Building the elbow jigs—Making the elbows proved to be much more complex than making the straight lengths of handrail. To shape the elbows I needed two jigs: one to form the outside curve and one to shape the inside curve (photos, p. 60). The jigs securely held each curved elbow blank in place for successive passes over the shaper knife. Most important, the jigs kept my fingers out of the picture.

Both jigs are similar in construction. A base rides against the lower ball-bearing shaper collar to guide the cuts. I veneered the underside of each base with plastic laminate to allow it to glide freely on the shaper table. The

Little in the foyer stair gets in the way of light flooding the hallway from skylights above. Tempered glass panels support a sinuous maple handrail, and open risers contribute to the stair's transparency. A second, similar stair was built elsewhere in the house.

front edge corresponds to the inside or out-side curve of the elbow, and determines the shape of the cut. Because the minor diameter of the cutter was offset by ¼ in. from the col-lar, the template curve and the cutting curve had to be slightly different. The purpose of the ¼-in. offset was to provide clearance so that the slightly oversize blank would not rub against the upper collar during the first two se-ries of passes.

Attached to each jig base is a layer of 5/4 poplar lumber. This layer lifts the workpiece so that the outermost portion of the shaper knife can reach under the work when cutting. This layer also holds captive the countersunk heads of carriage bolts; wingnuts and washers on the ends of the bolts put pressure on a caul that grips the workpiece securely.

Other elements of the jig's anatomy includ-ed a fence against which the workpiece would rest, stop blocks for registering the ends of the blanks and a pair of stout angle-iron handles for the outside-curve jig (the other jig could be

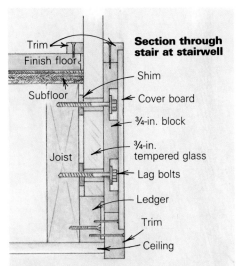

Blocks were dadoed to recess the lag bolts, leaving ¼ in. of wood to serve as a soft washer that cushions the glass. A ledger bears the weight of the glass; lag bolts had only to hold the glass in place.

used safely without handles). T-nuts embed-ded in the cauls held dog screws that helped to secure each workpiece. I filed sharp points on two of the dog screws used for the inside curve jig, figuring that this would give them extra purchase on the first two passes on the rough blank. Dog screws with a gentler 45° conical point were used on the third and fourth passes; they were positioned to exert a firm downward and inward pressure on the el-bow's finished surfaces.

To form each elbow, I'd have to make four series of passes: two for the inside curve and two for the outside curve (drawings, next page). Each series would form either an upper or a lower half of the elbow. In each series of passes the knife was raised in ⅛-in. incre-ments for successive passes until the full 1½-in. depth of cut was reached. Cutting a little at a time reduced the load on the shaper and produced a better finish in the final pass. Tearout, which would have been a real prob-lem with softer wood, was minimal with the

Inside- and outside-curve jigs

To form the inside curve of each railing elbow, McBride used the sturdy jig shown below. The base of the jig corresponds to the inside diameter of the elbow, and rides against the shaper collar. McBride tightened wingnuts against a caul to secure the workpiece; pointed dog screws added holding power. Slender, removable metal pins were used to shim out the rough stock when the first cuts were made. The pins were removed for subsequent cuts.

After slipping a partially milled elbow into the second jig and snugging up a battery of wingnuts against the caul (photo above), McBride made another series of cuts to form the outside curve of the elbow. Stout angle-iron handles allowed McBride to keep his hands well away from the cutting action. Conical dog screws helped to keep the elbow in place as it was cut. A stop block registered each elbow in the correct position.

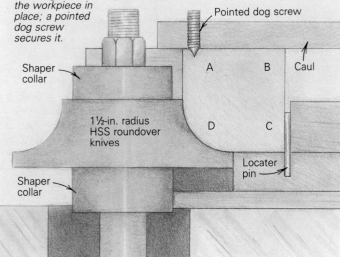

Inside-curve jig

First series of cuts: This series creates one half of an inside curve. The rough blank is registered against locater pins. The caul clamps the workpiece in place; a pointed dog screw secures it.

Pointed dog screw

Shaper collar

Caul

A   B

D   C

1½-in. radius HSS roundover knives

Shaper collar

Locater pin

Inside-curve jig

Third series of cuts: This series creates the second half of an inside curve. The outside of the elbow registers against stop plates.

Conical dog screw

Stop plate

D   C

A   B

Locater pin removed

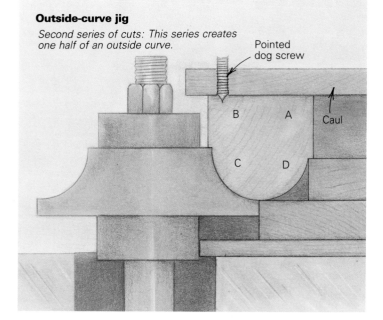

Outside-curve jig

Second series of cuts: This series creates one half of an outside curve.

Pointed dog screw

B   A

C   D

Caul

Outside-curve jig

Final series of cuts: This creates the second half of an outside curve.

Conical dog screw

C   D

B   A

close-grained maple, even on the short grain sections of the curve.

The inside curve—I began by thicknessing wide maple stock to 3 in. and bandsawing the blanks to rough shape, leaving an extra ⅛ in. on the inside and outside curves. The first series of passes would be made with the elbow locked into the inside-curve jig. For the initial cuts, I wouldn't have a true surface to work from, only the rough edges of the bandsawn blank. I'd need at least ⅛ in. of clearance between the fence and the elbow to accommodate the extra stock, and probably a bit more in places, so I made the fence to the elbow's outside radius of 5 in. and added ¼ in. to that. This offset provided clearance for the oversize blank, as did the previously described offset between cutter and collar.

To position the blank approximately in the inside curve jig for the first passes, I installed a pair of removable ⅛-in. dia. pins against the curve of the fence, and snugged the blank against the pins (drawings facing page).

The rest of the curve—After I cut the first half of the inside curve, the freshly milled surface could be registered against the convex fence of the outside-curve jig for the next series of cuts. The fence on this jig had been sized to the elbow's exact inside radius of 2 in. because the fence would be registering only previously milled surfaces. After locking the elbow into place, I made the second series of cuts, forming the lower outside curve of the elbow.

For the third series of cuts, I switched the elbow back to the inside-curve jig. Now I needed to reduce the radius of the inside-curve fence so it would register properly against the trued outside edge just cut in the second series. To do this, I slipped a pair of curved, ¼-in. plywood stop plates into the jig, just above the fence. The inside edges of each plate would ride against the outer circumference of the elbow, just above the midpoint of its thickness (drawings facing page). That way the plates wouldn't interfere with the as yet unmilled portion of the blank. For the fourth and final series of passes, it was back to the outside-curve jig. This series of passes corresponded to those made in the second series of passes.

Making accurate radial cuts—A 15-in. power miter box (Hitachi #C15FB) was used to cut the elbows and the handrails to length. Positioning the elbows for accurate radial cutting, though, took some thought.

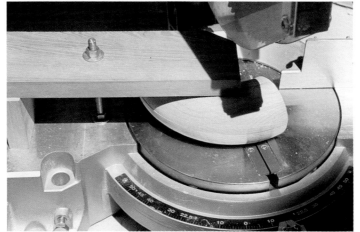

Straight lengths of handrail (top photo) were milled on a table saw fitted with a molding cutter. A guide strip mounted to the saw fence engaged a groove in the bottom of the handrail, thereby preventing the handrail from tipping as it moved past the cutter. Elbows were joined in combinations to make handrail easements, so the cuts between elbows (middle photo) had to be precise. To secure the elbows to his miter-saw table, the author drilled the table for a threaded stud and used it to secure a clamp block. To drill clearance holes for handrail bolts, the author used another jig (bottom photo). The drill bit was guided by a center hole drilled in a guide block of oak.

The Hitachi has a replaceable table insert with a slot the exact width of the blade kerf. I drew a pencil line on the table, running it across the faces of the split fence, and it crossed the slot at the imaginary center point. Measuring 5 in. out from this point along the left fence (the radius of the outside curve of an elbow), I drew a vertical line on the fence. By positioning the outside circumference of an elbow against the line while but-ting its end tight to the fence, I would ensure a radial cut. Because the pivot point of the miter box lies behind the fence, however, the center point would shift slightly from left to right when cutting different angles. I adjusted the vertical line on the fence for the different cuts, in each case measuring 5 inches from wherever the plane of the fence crossed the cutting slot.

To hold the elbows securely while cutting them, I drilled the miter-box table for a threaded stud. A clamp block and nut atop the stud left my hands clear of the action (middle photo). With a 100-tooth carbide blade on the saw, the result was a cut smooth as glass.

Assembling the handrail—To join the handrails to the elbows, I used double-ended handrail bolts, which have a machine thread on one end and a lag thread on the other. To connect elbows to each other, I used dowel screws; they have a lag thread on both ends.

To drill accurate pilot holes in the elbows and the railing, I built drilling jigs. These were just squared-up blocks of oak, drilled on a drill press and fastened to one of two bases. One base supported straight lengths of rail that were secured by a pipe vise (bottom photo). The other was for elbows, using a rod-and-nut clamping arrangement similar to the one used earlier for the miter box. The clearance hole in the handrail was drilled oversize to provide some margin for error in aligning the elbow to the handrail. A nut pocket mortised into the underside of the handrails intercepted the clearance hole, giving me just enough room to slip a nut onto the machine screw and tighten it with a box wrench.

The handrail bolts made a rock-solid joint as far as keeping the parts from separating, but I was worried about misalignment of the joints over the years. To ensure that this would not occur, I keyed the parts together after assembly using a plate joiner. Biscuits were let halfway into the handrail surface in inconspicuous locations. After the glue dried, the biscuits were shaved down flush and the joints were faired with planes, rasps, and sandpaper. Finally, the gap between the glass and the outside edges of the glass groove was filled with silicone caulk, forming a bond between the glass and the handrail. □

Scott McBride is a contributing editor of Fine Homebuilding. *Photos on this page by the author.*

Cantilevered Stairway

Steel brackets support cantilevered treads around a cylinder

by Rob Harlan

Curves make my heart sing. When I built my own house, one of the things I did was to frame a 16-ft. diameter, two-story cylindrical room. The curving wall was a great place to locate the stairway (photo below). I had a picture of a curved adobe building with slabs of wood for stair treads cantilevered out through the wall. I wanted to build a similar stair, but the problem was how to capture the adobe-house stair's light, suspended feeling, yet have it structurally sound and in compliance with the building code.

Steel solves the problem—I tried a whole bunch of wood-based designs, but the only one that worked involved extending the cantilevers through the other side of the wall, and that created space problems. Then, a friend suggested using steel to support the stair treads. Being primarily a woodworker, I had to be talked into it, but steel made for strong cantilevers that I could contain within the 2x6 wall.

I had a machinist make stair-support brackets from ¼-in. by 4-in. plate steel. Most of the brackets are two plates lapped and welded to form a T-shape. The top of the T is fastened to the studs with four ½-in. machine bolts; the base of the T protrudes through the curved wall and supports

No visible means of support. Projecting from a curved wall, this custom ash stair is supported by cantilevered steel brackets housed in the tread supports. The treads are glued and screwed to the supports, and the balustrade of bent laminations not only enhances the organic look but also helps stiffen the stair.

the treads (top photo, right). An engineer OK'd the bracket design, and the building department approved the stair.

The brackets are not all T-shaped. The bottom two are L-shaped, and the top ones are attached to the second-story floor joists, so they're straight.

One stud per tread—The cylindrical room was framed with 2x6s on 16-in. centers. In the area where the stairs project from the curved wall, I used doubled 2x6s to make 3-in. by 5½-in. studs on 9 ³/₁₆-in. centers at the outside edge. Each of these studs bears a stair bracket.

The stud spacing is based on some fairly simple math. The stairs are laid out with a 6¹⁵/₁₆-in. rise and an 11-in. run. Because the treads are wedge shaped, the run is longer out at the railing than at the wall. The 11-in. run is at the center of the tread, where one is most likely to walk. All I had to do was figure out the run at the wall.

I knew the radius of the circle to the outer edge of the stud wall (96 in.). I also knew the radius of the circle at the center of the treads (115¼ in.). I knew the run of the stairs at the center of the treads (11 in.). I determined stud spacing with a ratio: 115.25 ÷ 11 = 96 ÷ x. The radius and the run at the center of the treads are directly proportional to the radius and the run at the wall. Putting in the numbers yielded 9³/₁₆-in. o. c. for the stud spacing.

The double studs were marked at the correct heights, using a builder's level and a story pole, and then drilled out for the bolts that secure the steel brackets to the studs. The lapped portions of the brackets were let into the studs so that the brackets could be bolted tight to the studs.

After walking up and down the supports a few times to verify their soundness and checking the layout for consistency of spacing, rise and levelness, I unbolted the brackets, rubbed them with stove black to prevent rust and stored them.

As construction progressed, I wrapped the outside of the curved wall with ⅜-in. CDX plywood and later with ½-in. drywall (middle photo, right). Slits were cut through these wall coverings, and eventually the steel supports were reinstalled from the back and bolted into position permanently before the drywall on the other side of the wall was installed.

Supporting the treads—I chose white ash for the entire stairway—the tread supports, the treads and the balustrade—because of its light color, durability and bendability.

Each tread support consists of three pieces of ⁵/₄-in. by 6-in. ash sandwiched over the steel bracket. The outer section of the tread support was mortised with a router to create a pocket to house the steel (bottom photo, right). The tread support was glued with Tightbond ES747 and screwed to the steel, which had been drilled with eight ⅛-in. countersunk holes.

The remaining two sections of the tread supports were glued with Weldwood urea resin and screwed to the outer piece, covering the steel. Great care was taken to assemble the tread supports around the steel so that the joints were flush and to clean up all glue drippage with water immediately to minimize sanding after assembly.

Brackets fit in the wall. The doubled 2x6 studs were laid out on 9 ³/₁₆-in. centers—the run of the stairs at the wall—so that a welded-steel bracket could be bolted to each stud.

Drywall went on first. Slits were cut in the plywood-and-drywall sheathing materials so that the brackets could be reinstalled from inside the cylindrical room.

Tread supports sandwich the brackets. The three-piece ash tread support was mortised and fastened to the steel with Tightbond ES747 construction adhesive and screws. Weldwood urea resin glue and screws bonded the tread support, which provides bearing for the treads.

I tapered the tread supports with a bandsaw, then planed, routed, sanded and installed them. Tapering gives the suspended staircase the light look I wanted.

Although the steel brackets cantilever through the wall, the wooden tread supports were installed on top of the finished wall, which provided for easier plastering and painting. It also kept any movement of the stairs from cracking the plaster. To give the stairway its "coming out of the wall" look, I caulked where the tread supports met the wall with flexible caulk and then carefully painted it.

Making treads—The treads are ⁵/₄-in. by 12-in. pie-shaped pieces blind-splined together from 6-in. wide stock. The splines are ⅛-in. by 1½-in. pieces of white ash placed in ⅛-in. slots in the treads. Both the splines and the slots were made with a table saw.

The treads were glued and screwed to the tread supports. I left a uniform ¼-in. gap between the finished wall and the end of the tread to heighten the floating feeling.

Once again, I countersunk the screws and plugged the holes. The landing was made similarly by gluing and splining many pie-shaped pieces together.

Laminated balustrade—The balustrade I designed for this stairway preserves the light, suspended feeling of the cantilevers and weaves them together, which stiffens them. Each baluster consists of a pair of bent laminations. I made two separate bending forms, one for each baluster shape, then glued up twelve ³/₃₂-in. wide strips of ash and clamped them in the forms.

Once the glue had cured, the bent laminations were trued up using a table saw to clean and cut the edges square, then finished with a belt sander and rounded over with a router and lots of hand sanding. Next, the balusters were glued, screwed and plugged together at their bases to form pairs.

The handrail was laminated from eight ⁵/₁₆-in. by 1¼-in. by 16-ft. long strips of ash. The problem was figuring out the radius of the handrail and how to bend it.

The process turned out to be easy. I didn't have to use math; I laminated the rail in place using the outside ends of the treads as a form to hold the railing in the correct bend. The strips were clamped together with C-clamps every 3 in. It's moments such as these when you use all of your clamps and any that you can borrow.

Once the railing lamination was sanded and routed to its final shape, I set it temporarily in place. The baluster assemblies were carefully marked and cut to fit below it. I attached the balusters to the railing with ¼-in. by 2-in. dowels. The base of each baluster was mortised into the tread and glued and screwed to it.

When the balustrade was complete, the entire structure stiffened up considerably, and my dream of an organic, strong, light, curved stairway was achieved. ☐

Rob Harlan is a licensed general and solar contractor in Mendocino, Calif. Photos by John Birchard.

Building an L-Shaped Stair

Basic construction techniques for a stair with a three-step winder

by Larry Haun

Thirty-five years ago, orange blossoms scented the clear southern California air, streetcars hadn't yet surrendered to freeways, and the once-arid San Fernando Valley just north of Los Angeles was about to be transformed into suburbia by armies of construction workers. As the demand for housing increased almost daily, carpenters developed new techniques to keep up with demand.

A project I worked on back then was a 60-unit apartment house. Each unit had its own L-shaped stairway to the second floor, and each stairway had a three-step winder in it (a winder uses a series of wedge-shaped steps to make a stairway turn 90°). Not knowing how to construct a winder, I relied on my carpentry books for information. Using that information on the job the next day, my partner and I cut and built one set of stairs, but the winders alone took us more than five hours. Now, faster certainly does not always mean better, but that was simply too long. So we set out to find a winder technique that would combine speed with the quality of traditional carpentry techniques. The method we now use is quite simple and results in a stair strong enough to withstand the rigors of long service.

Merit and demerit—An advantage of winders in a stair is that they shorten the total run, thereby leaving more space at the top or bottom of the stairs (drawing facing page). This is particularly helpful in small homes with limited floor space. The majority of winders I've seen are carpeted later on, but I have also seen many winders covered with hardwoods, even tiles, that look quite beautiful.

Whatever the surface, however, a stair with winders is not without danger. The fact that the width of each tread in a winder varies can present a hazard, unless you take care to step where there's enough tread width to support you. Building codes regulate the shape of treads in winders, though the codes vary from state to state. Some allow a winder to come to a

Traditional winder construction methods apply treads and risers to short stringers. Production methods offer a quicker approach.

point on one end of the tread. More often, however, codes require the tread be 6 in. wide at the narrow end, or have a 9-in. to 10-in. wide tread at the "line of travel," which is the path a person would likely follow when ascending or descending the stairs (drawing facing page). The line of travel is generally 12 in. to 16 in. away from the narrowest portion of the stairs.

Any stair should feature properly sized and secured handrails, but good handrails are particularly important on stairs with winders (for additional information on proper handrail design, see "Design Guidelines for Safe Stairs," pp. 15-17).

First things first—When building a set of stairs with a three-step winder, begin construction as you would with most any set of

stairs. First, determine the total rise (the vertical distance from the first floor to the second floor). Divide the total rise by 7 in., a comfortable and safe riser height for most people. Round this number off to get the number of risers needed in a flight of stairs. Once you know the number of risers, divide that number into the total rise. This will give you the exact rise of each step.

Here's an example. Say the total rise from first to second floor is 108 inches. Divide 108 by 7 and you'll get 15.4. Drop the fraction (.4) because it isn't safe to have a part of a step in a flight of stairs. Next, divide 15 risers into the total rise (108 divided by 15 equals 7.2). So in this particular flight of stairs, there will be 15 risers, and each riser will be 7.2 in. high (7³⁄₁₆ in.).

Building a platform—Construction of the winders is straightforward and as easy as building boxes. The first step is to build the platform. Say that you are building a set of stairs in which the rise for each step is 7³⁄₁₆ in., the winder begins at the fourth step, and that the stairs are 36 in. wide. Frame a 36-in. square platform for the stairs and deck it off (before nailing down the plywood, I run a bead of construction adhesive along the top of the framing as a squeak stopper). The platform should be 28³⁄₄ in. high (four risers with a 7³⁄₁₆-in. rise equal 28³⁄₄ in.—simple enough).

Laying out the winders—Traditionally, carpenters have constructed winders by cutting out stringers, much like they do when building a regular set of stairs. This method is familiar across the country, and many of those staircases are works of skill and art. Production framers, on the other hand, eliminate winder stringers altogether; they simply build three boxes and stack them one on top of another. This is much easier than the traditional method and saves a considerable amount of time. The width of most main stairways is 36

Drawings: Bob Goodfellow

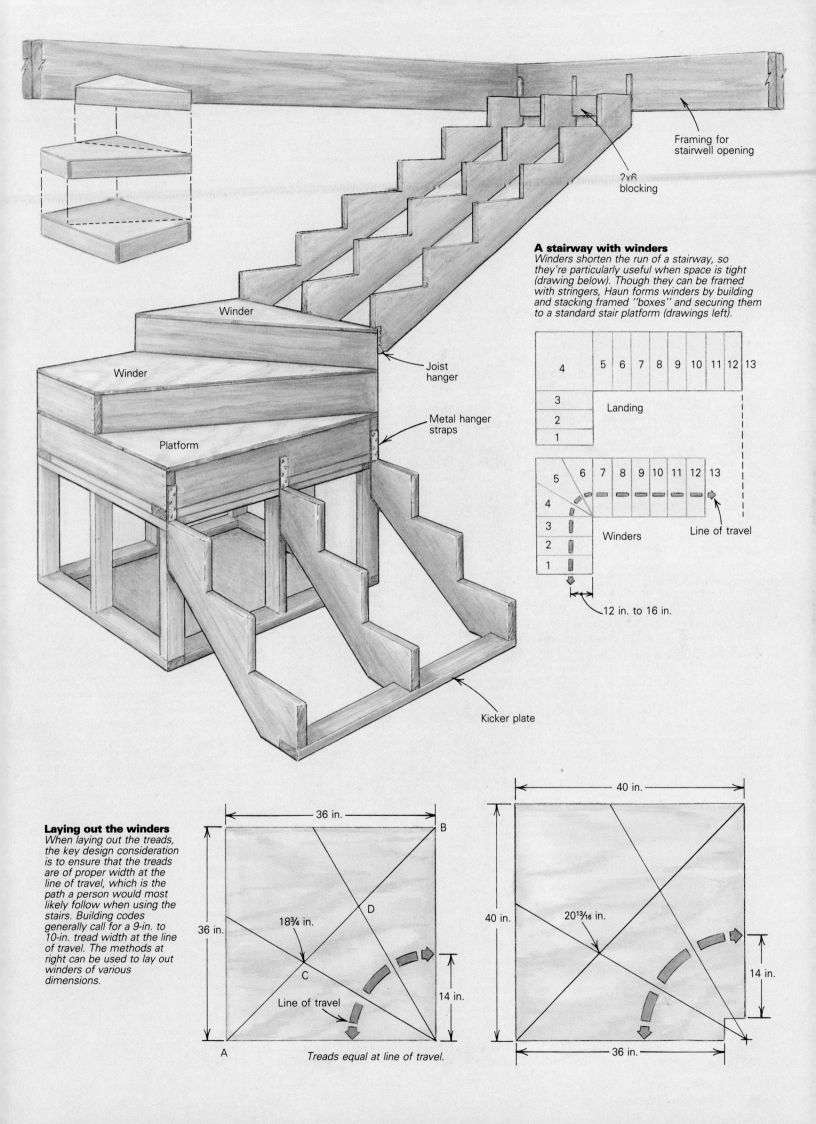

Framing for
stairwell opening

2x6
blocking

Joist
hanger

Metal hanger
straps

Winder

Winder

Platform

Kicker plate

A stairway with winders

Winders shorten the run of a stairway, so they're particularly useful when space is tight (drawing below). Though they can be framed with stringers, Haun forms winders by building and stacking framed ''boxes'' and securing them to a standard stair platform (drawings left).

4		5	6	7	8	9	10	11	12	13
3		Landing								
2										
1										

Winders

Line of travel

12 in. to 16 in.

Laying out the winders

When laying out the treads, the key design consideration is to ensure that the treads are of proper width at the line of travel, which is the path a person would most likely follow when using the stairs. Building codes generally call for a 9-in. to 10-in. tread width at the line of travel. The methods at right can be used to lay out winders of various dimensions.

36 in.

36 in.

B

D

18¾ in.

C

14 in.

Line of travel

A

Treads equal at line of travel.

40 in.

40 in.

20¹³⁄₁₆ in.

14 in.

36 in.

in., so I'll assume that width in the examples to follow.

Start with a 36-in. square of ¾-in. plywood, and divide it into treads so the line of travel for each tread is the same. You can make the division using the "hit and miss" method, or you can use a more technical approach.

My hit and miss way begins with a line snapped across the plywood from corner to corner, from point A to point B (bottom left drawings, previous page). Measure 18¾ in. on the diagonal from Point A to find Point C. Do the same from corner B to find Point D. Where did the 18¾ in. come from? After trying lots of dimensions, I've found that this one works to make a comfortable stair. Next, snap two lines: one through Point C and the other through Point D. These lines define the three treads of the winder.

The other method of laying out winders is more versatile and requires some arithmetic, but not much. Take the 36-in. square piece of plywood and divide it diagonally as before. To find Points C1 and D1, multiply the length of the plywood by .52 (36 x .52 = 18.7). In this example, therefore, the distance in from each corner of a 36-in. square is 18.7 in., or about 18¾ in. Snap lines through the points you find to determine the winder layout. You can use this formula to figure out the winder layout of any size square or any width of stair. For example, if the stair width is 30 in., you would need to measure in 15.6 in. from each corner (30 x .52 = 15.6).

As I noted earlier, building codes frequently call for a winder tread with a width of at least 9 in. or 10 in. at the line of travel. If this is the case, you'll need a 40-in. square piece of plywood to make the platform for a 36-in. wide flight of stairs. You'll have to project the inside corner point far enough out diagonally so there is enough tread width at the line of travel.

Establish the 36-in. width of the stairs on the plywood as shown (bottom right drawing, previous page). Then proceed to lay out the line-of-travel radius and divide it into three equal arcs. Connect points on the arcs with lines leading to the inner corner of the layout to complete the winder layout. This will provide a 10½-in. tread at a 14-in. line of travel.

Constructing the risers—After laying out the risers on a sheet of plywood, cut them out with a circular saw. The rise of the step in our example is 7³⁄₁₆ in., so take some 2x8 stock and rip it down to 6⁷⁄₁₆ in. You might as well rip enough to do all of the risers—you'll need about 24 lineal feet in all. The ripped stock will form the actual riser of each step, as well as the "joists" that support successive steps. Once you nail the 2x stock together, sheet the frame with the plywood. Toenail this box to the landing.

Now take the smallest piece of plywood, the one shaped like a piece of pie, and build the last riser with the last of the ripped-down 2x8 stock. Toenail this box on top of the previous one. That's all there is to it—the landing is now a three-step winder (bottom photo, right).

Stacking steps. The winders are simply triangular or trapezoidal boxes made of 2x framing; plywood forms the steps (photos below). After the stair platform has been framed, the first box is toenailed to it, after which the second box is toenailed to the first.

Installing stair stringers—With the landing and winders complete, you can cut and attach the stringers for the rest of the stair. These stringers, both upper and lower, are laid out and cut as usual (for additional information on laying out and cutting stringers, see pp. 8-14). A stringer from the first floor to the landing requires three risers; the fourth riser is created by the landing itself when the stringer is hung from it. Stairs that are 36 in. wide require three stringers—one on each side of the landing and one in the middle—and are usually cut from 2x12 stock.

One way to hang the stringers is to nail an 18-in. long perforated metal strap to the bottom edge of the stringers. This should be a heavy-gauge strap, not plumber's tape or the like. Nail the last 6 inches of the strap to the bottom of the stringer with four 16d nails. Then bend the strap around the backside of the top riser either by hand or with a hammer. The strap should extend above the top tread by about 6 inches.

To locate the stringers on the landing, measure down 7³⁄₁₆ in. from the top of the joist and strike a line parallel to the edge. Hold each stringer to the mark, and secure it by nailing the end of the metal strap into the face of the platform with four 16d nails. The straps carry the actual weight of the stairs.

Next, nail a pair of 2x6 blocks (in this part of the country, we call them pressure blocks) to the platform and between the stringers. This block helps keep all three stringers stable until the stairs get their treads and risers. Secure the bottom of the lower stringers with a 2x4 kicker plate. This plate should be as long as the finished stairs are wide. Slip it into a notch at the bottom of the stringer and nail it to the floor.

Another way to secure the lower set of stringers is to use joist hangers. Using a circular saw, kerf the back side of the last riser. Slip the joist hanger into the saw cut and nail it to the joist and then to the landing.

After installing the lower stringers, you can install the stringers that run from the landing to the second floor. The top of each stringer is attached to the second-floor framing with metal straps, just as the lower stringers were secured to the landing. The bottom of each stringer is secured to the landing with a joist hanger. I cut 1½ in. off the bottom tip of each stringer, which allows it to bear nicely on a joist hanger. If the inside stringers are against a wall, we nail a Simpson A35 framing anchor at the juncture of the stringer and the platform.

Securing stringers in this fashion is a tried-and-true procedure; we have been doing it for more than 30 years. The stairs will not come loose no matter how much the wood shrinks and no matter how heavy the load they're asked to support. □

Larry Haun lives in Los Angeles and is a member of Local 409; he was a longtime teacher in the apprenticeship program. Photos by Roger Turk.

A Hollow-Post Spiral Stair

If you keep it simple, you don't need much more than a table saw and a drill to build one

by Peter Lucchesi, III

I bought my house in 1969. At that time it was a summer camp, built into a steep hill that plunged into the Connecticut River. For ten years I lived in what was essentially a Hobbit hole.

My remodeling project began, innocently enough, as an attempt to fix the leaks that inevitably occur on flat roofs in New England. Before I was done, though, I built a whole new roof and a 900-sq. ft. addition. The biggest challenge of the project proved to be building the spiral staircase up to the bedroom loft (photo below).

Given the space limitations inside the addition, a spiral stairway was the obvious choice for access to the loft. But how to build it and where to put it were not so obvious. I ended up centering the stair under the apex of the cathedral ceiling.

Material selection for the stair came down to a choice between cherry and oak, since both were available locally at about the same price. I chose oak because of its light color, thinking it would contribute to the light and open feeling I was trying to incorporate into my former Hobbit hole.

Scaling the heights—The distance from finished floor to finished floor measured 94 in. Eight inches is usually the maximum rise allowed by code on a straight run of stairs. But most codes permit a steeper rise on spirals (up to 9½ in.). I chose 8 in. as a starting point and divided it into my total rise of 94 in. The result was 11¾. I rounded that off to 12, meaning 12 rises over 94 in. Dividing 94 by 12, I came up with an individual rise of about 7⅞ in. for each step. My stair would have 12 risers and 11 treads, since the number of treads is always one less than the number of risers.

Post time—I'm not a mathematician and had never built a spiral stairway before, so most of the work was done by trial and error. I figured it would be tough to make a perfectly round post, and besides, I wanted something I could make on a table saw. Also I wanted the post to be hollow, with access to the inside so I could bolt the treads in place. I decided to make the post an octagon. I'm not sure why I chose eight sides, instead of six or ten, but I liked its shape and it seemed like it would work right for the tread layout.

I had the local millwork shop, Amherst Woodworking and Supply, mill the lumber for the sides of the post into 1¾ in. by 3 in. by 12 ft. lengths. On their advice, I stacked and clamped the pieces together until I was ready to cut them, which helped keep them from twisting out of shape.

Before cutting the oak, I experimented with spruce 2x4s, ripping them to 3-in. widths and running them through the table saw set at 22½°. I found that I had to adjust the angle a few times before the last piece would fit together with the others.

Once I got the angle right, I cut the bevels on the oak—a very slow and tedious process. When all eight pieces were beveled, I clamped them together again; I didn't want to take a chance that they might twist overnight.

The next step was to assemble the post. I cut a half-dozen octagonal blocks out of 1½-in. thick oak to fit inside the post and located them where they wouldn't interfere with bolting the treads later (drawing on following page). I assembled the whole thing as a unit, gluing the sides of the post to the blocks (and to each other) with yellow glue. I left the glue off the last side so it could be

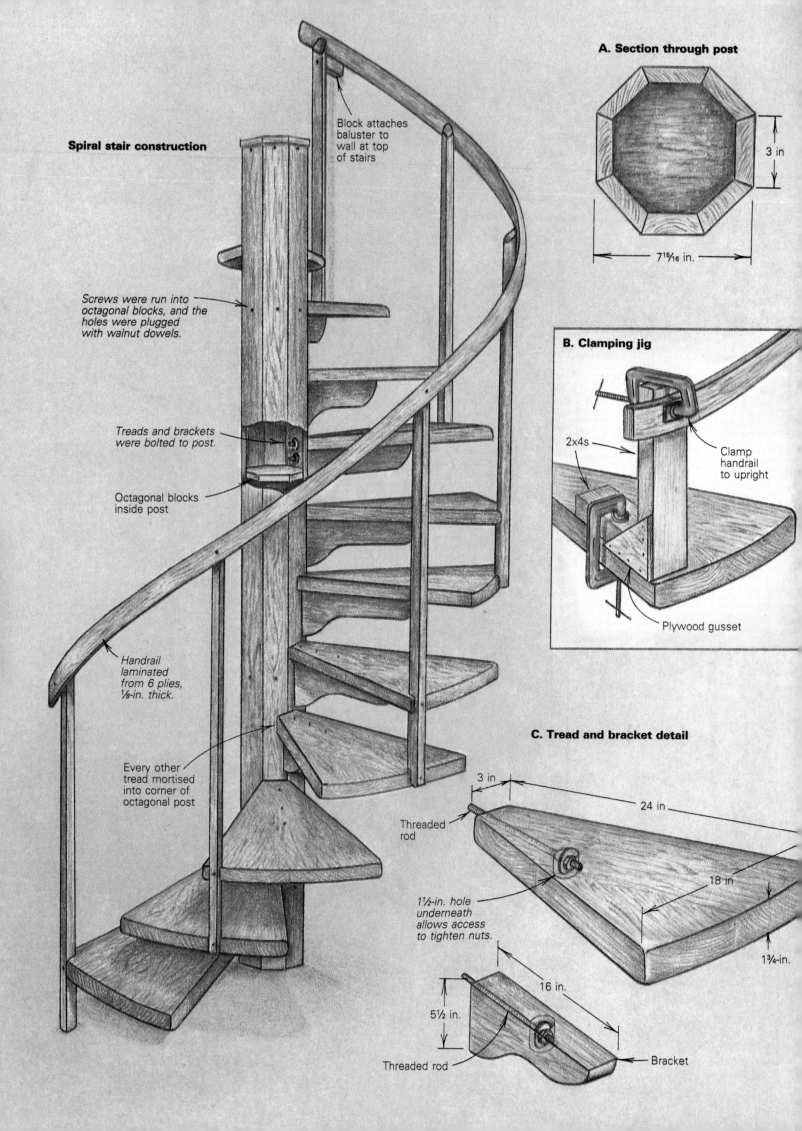

Spiral stair construction

Block attaches baluster to wall at top of stairs

Screws were run into octagonal blocks, and the holes were plugged with walnut dowels.

Treads and brackets were bolted to post.

Octagonal blocks inside post

Handrail laminated from 6 plies, 1/8-in. thick.

Every other tread mortised into corner of octagonal post

A. Section through post

3 in

7¹⁵/₁₆ in.

B. Clamping jig

2x4s

Clamp handrail to upright

Plywood gusset

C. Tread and bracket detail

3 in

24 in

Threaded rod

18 in

1½-in. hole underneath allows access to tighten nuts.

1¾-in.

5½ in.

16 in.

Threaded rod

Bracket

removed later and clamped the whole affair with perforated metal strapping (also called plumber's tape) held together with bolts.

After the glue had dried and the clamps were removed, I pulled off the unglued side and put it out of harm's way. Then I ran 3-in. #8 screws into all of the blocks through counterbored holes in each of the sides and plugged them with walnut. If my glue joints were perfect, then the screws were an unnecessary precaution, but I felt better knowing they were there, and too, the walnut plugs later became a design theme. The finished post is just under 8 in. in diameter at the widest point.

Treads—I made plywood patterns for the treads and brackets. Amherst also glued up and cut out these pieces for me. The treads are 1¾ in. thick and 24 in. long. Most codes require that treads be at least 26 in. long on a spiral stair, but local codes may vary and the final decision always rests with the building inspector. Treads taper from 18 in. wide on the outside to 3 in. wide where they butt into the post. Most codes require that the treads on a spiral stair measure at least 7½-in. wide at a point 12 in. from the inside edge; mine measured 10½ in. at that point. The brackets are 16 in. long and 5½-in. deep.

Using a 24-in. long twist bit chucked into a portable drill, I drilled a ½-in. hole 10 in. deep down the length of each tread and bracket (detail C, drawing facing page). On the underside of each tread and on the top of each bracket, I then drilled a 1½-in. hole deep enough to meet the end of the ½-in. hole (the same way you would when connecting handrail parts together with a railbolt). A Forstner bit would have been a good choice for this because it drills a flat-bottomed hole. But all I had was a spade bit, so I watched the depth carefully to make sure the center point didn't come through the top of the tread. I ran ½-in. threaded rods through the treads and brackets, leaving about 2 in. sticking out. I secured them with nuts and washers located in the 1½-in. holes, which were large enough to get a box wrench on the nuts and tighten them.

Next I screwed and glued the treads and brackets together. I ran three 2-in. screws down through the top of the treads into the brackets, and once again, counterbored and plugged the holes with walnut.

I stood the post in place temporarily and marked the locations for the tops of the treads. Then I took the post down and laid it on a pair saw horses, ready for drilling.

The alignment of the threaded rods was a little different on each tread and bracket. So I measured each set of rods with a pair of dividers and transferred the measurements to the post. I drilled the holes in the post freehand, but held a square beside the bit to help me keep it perpendicular.

False start—I stood the post in position, screwed it to the subfloor through a block in the bottom of the post and braced it tempo-

The treads and brackets were screwed and glued together, then bolted through the central post. Each baluster was screwed to two treads—mortised into the upper one and scribed over the lower. *Photo by Wendy Jackson.*

rarily from above. Then I began attaching the tread and bracket units, bolting one to each side of the post. I glued all surfaces that made contact with the post with yellow glue and, taking care to wipe off any excess glue with a wet sponge, I tightened the nuts with a ratchet as I went. I found that by leaving out one of the post's staves, I was able to get a ratchet on all the nuts.

I had six treads attached when I realized that my plan wasn't going to work. By the time I got around to the seventh tread, there wouldn't be enough headroom beneath it to get on the first step (unless I was 4 ft. tall).

Since I had paid over $850 for all the lumber and millwork, I wasn't about to scrap anything. I decided to remove the treads and brackets, patch the holes and turn the post end for end.

When I undid the nuts and set out to break the glue joint holding the treads in place, I discovered something that I'd long suspected—glue does indeed hold to end grain. I was able to stand on the treads without breaking the glue joints. I really had to whack them with a framing hammer and a block of wood before they came apart. And even then, shreds of the post came off, stuck to the end grain of the treads.

I started again, and this time, I mounted the first tread on one side of the octagon and mortised the next into the point (or corner) between the first and second side.

With all the treads attached, the stair made only about a 220° revolution, instead of a full 360°. It is somewhat steep, but very manageable. I sanded it down and put a quick coat of urethane on the treads, brackets and the post, because I didn't want any moisture getting in to make the wood expand.

Since wood compresses, I tightened each nut inside the post every couple of days for the next few weeks. When I was satisfied that the wood had compressed all it was going to, I glued the last stave of the post in place.

On the rail—I used a very simple post and rail design for the balustrade. To make the handrail, I ripped six pieces of oak ⅛-in. thick and 2½-in. wide on the table saw. Then I made clamping jigs using two short lengths of 2x4 joined at right angles and strengthened with plywood gussets. I C-clamped one of these jigs to each of the stair treads and used them as forms for the handrail (see detail B, facing page). Gluing up two plies at once was all I could handle in the time it took to get the glue spread and the clamps turned down. I needed more clamps than I had, so I made some with ¾-in. pine and drywall screws. The handrail had to be clamped about every three or four inches.

Once all the plies were glued up, I removed the clamps and went at the handrail with a belt sander. Since there was a lot of unevenness, it took several hours of sanding.

I made the balusters out of 1½-in. oak and radiused the edges with a ½-in. roundover router bit. For stronger support, I attached each baluster to two treads, mortising them into the first and scribing them over the second (photo above). I glued and screwed them in place, and of course, covered the screw holes with walnut plugs. The handrail is simply screwed to the side of the balusters and braced to the wall at the top of the stairs. □

Peter Lucchesi, III is a carpenter and builder in South Hadley, Mass.

Octagonal Spiral Stairs

A complicated stairway built and installed in separate halves

by Tom Dahlke

The house Gary Therrien built for himself is octagonal, with a truss-roof system that required no load-bearing interior walls. The floor of the large, open main level emphasizes the shape of the house: oak boards laid to form wedge-shaped sections that were edged with mahogany strips running to the house's corners. In the center of the floor is yet another octagon: the opening for the staircase that leads down to a lower-level bedroom, the entry, workshop and playroom. Therrien wanted a spiral stairway, but the curves of the standard round or oval spiral wouldn't work with the angles and straight lines of the rest of the house. He felt, and I agreed, that it would be more appropriate to build a stair whose outside edge followed the shape of the opening in the floor. The fact that this octagon didn't have equal sides was something I didn't want to think about yet.

The basic spiral staircase isn't all that hard to lay out or build, but I was dreading the thought of putting together the unsupported octagonal stringer for this one. I also realized early on that the walls around the opening on the bottom floor would make it impossible to stuff a fully assembled unit in place, I would have to build the stairway in two vertical halves and install them one at a time.

The opening was elongated—6 ft. 6 in. wide by 7 ft. 6 in. long. Each end had three 31-in. sides joined by two 43-in. sides. The stair treads had to spiral 360° over a total drop of 9 ft. 3 in. I decided on 14 treads with a rise of 7⅜ in., and began by drawing up a plan view

The opening for this stairway was an elongated octagon in the center of an octagonal house. The floor on the main level was laid in wedges to emphasize the unusual shape, and the owner wanted an octagonal spiral staircase to carry out the idea. The completed stair makes a full 360° spiral. Its unsupported stringer describes an eight-sided figure that's echoed by the rail.

Assembling the post. The two halves of the post were coopered up separately. Some of the blocking inside is reinforcement, right, and some will support the octagonal blocks that the treads will be bolted to. The blocks also help align the halves when they are assembled. Mortises, each one different, have already been cut for the treads. Above, the post is strapped together dry, before the treads are installed.

Each tread passes through a mortise in the post and is lag bolted in place through the octagonal bolting block.

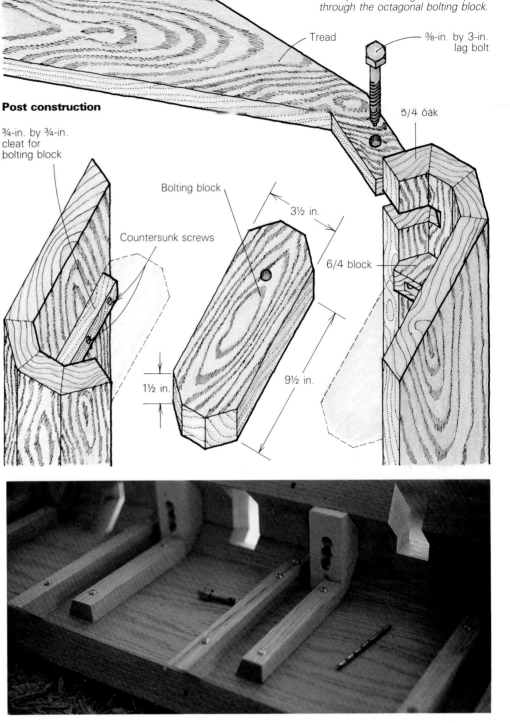

Post construction

¾-in. by ¾-in. cleat for bolting block

Tread

⅜-in. by 3-in. lag bolt

5/4 oak

Bolting block

Countersunk screws

3½ in.

6/4 block

1½ in.

9½ in.

and two elevations that showed where each tread would join the post and at what angle. The treads were laid out at 22½° intervals.

I set up my table saw (which has a mortising attachment), my jointer-planer, and my bandsaw in the living room, and did all the work on site.

The hollow post—I planned to rabbet the treads into the outside stringer. Since most of the stringer would be unsupported by anything else, the treads had to be locked immovably to the central post. Because of this, and because of the space problem, I decided to cooper up a hollow post. This would let me fasten the treads very sturdily on the inside. It also made it fairly simple to assemble the stairway. I would glue up two vertical halves

of the post, and bolt the treads to 1½-in. blocks shaped to fit inside the hollow space (drawing, above). These bolting blocks would rest on cleats screwed into the post. Other bracing would serve as guides when I glued and screwed the post halves together.

I used eight boards 9¼ ft. long and 1¼ in. thick to make the hollow octagonal post. To get the correct sectional dimensions, I ripped two of them to a width of 8½ in. and six to a width of 2½ in. Next, I tilted the blade on my table saw to an angle of 22½° and beveled the two edges of each piece. Then I put them together dry with strap clamps while I marked the post for the mortises. I had to scribe each mortise on both the outside and inside of the post to determine the angles at which the treads were coming in. To do this, I marked

their locations in plan view on one end of the post, and measured down the inside and outside surfaces. I also drilled holes every 15 in. along the edges of the two halves of the post. These were for the countersunk screws that would hold the pole together after assembly. Then I took the strap clamps off, and cut the mortises on each half with the bandsaw where possible, or with handsaws and chisels. No two mortises were alike, and I had to cut them precisely so the treads would mate with the stringer. When all the mortises were cut, I strapped the halves together dry again to have a look (photo above left).

Treads and stringer—No two treads are alike, either. Their outlines depend on the angles at which they meet the post, and the

Photos facing page: Donna Coveney; Illustration: Vince Babak

The treads were installed while the two halves of the post lay across sawhorses, top. The stringer sections, which had previously been routed out, were temporarily attached to the treads, and then carefully trial-fitted, above, to achieve a smooth joint between sections.

shapes imposed on their outer ends by the rising and angling stringer. I cut the treads long enough to fit into ½-in. deep mortises in the stringer, then made sure each one fit the post. Finally, I rounded their edges with a router and finish-sanded them with 220-grit paper. With the post halves resting on sawhorses in the cellar, I fit the treads into their respective mortises and bolted them to the interior blocks with 3-in. lag bolts, as shown in the photo at left.

Treads in place, I began cutting the stringer to fit their ends, and from here on, my drawings were worthless. Each joint required a compound-angle cut. I knew that all the miters had to be 22½° (actually 67½°, but for setting the blade angle on a table saw, whose protractor reads from 0° to 45°, you have to use the complement of 67½°, which is 22½°) for the stairway to make 135° turns, but the vertical angles varied because the rise of the segments varied. (The segments of the octagon were not equal, and this was the only way to be sure that the front and back edges of the treads would remain within the width of the stringer.) I'm sure there is a way to figure this stringer out on paper but I could only get part of it, and wound up trial-fitting instead.

I used three 12-ft. oak 2x12s for the stringers, with the best one in the middle where it is most visible. I rough-cut the segments to length one after the other down a board so the grain would be continuous on both sides of a joint. Then, with the post halves bristling with treads still on the sawhorses, I used my drawings and some simple calculations to lay out the mortises on the stringers that would house the wide ends of the treads. I routed out the mortises and set the rough-cut stringers in position on the treads. I predrilled each segment so it could be bolted to the treads with 3½-in. lag bolts countersunk and plugged with 1-in. oak plugs.

From here on, I could concentrate on the joints between stringer segments. I started at the bottom and worked up, cutting and recutting each segment until all the joints were tight (photo bottom left). I predrilled them for six 2-in. screws, to be countersunk and plugged with ½-in. oak plugs. Then I beveled the stringers' edges and used my table-saw mortising jig to drill dowel holes for the rail balusters. Finally, I sanded the sections with 220-grit paper.

Installation—On the day of the raising we had ten people on site, and we needed every one of them to muscle the two halves of the stairway into place. The stringer segments that could be installed later were left off, but the segments at the top had to be installed before the stair could be positioned. At the third and fourth steps from the top, the stringer passes within ¼ in. of a wall, and after we'd maneuvered the first half in where it belonged, we lag-bolted through the stringer segment into the studs. Two 1-in. dia. oak pins through the bottom octagonal bolting block hold the post in place on the floor.

With one half secured, we slid the other into place to check the fit one last time. It mated perfectly, so we glued, clamped and screwed the halves together.

After the remaining stringer segments were glued, screwed and bolted in place, I plugged all holes with oak plugs and sanded them smooth. The bottom section was drilled to sit on a 1-in. dia. oak peg set in the flooring.

The top of the post is held in place by a landing support cut from the stringer stock. A one-of-a-kind joint was cut so the 2x12 oak would join the post flush. To continue the effect of the spiral, we built the landing in two tread-size sections to give the illusion of two more steps.

Bannister and balustrade—The owner wanted the bannister to look like a continuous, floating ribbon, beginning over the post, going around the opening, then down and around the stair. I bought five rough-cut 14-ft. full 2x6 mahogany boards, then resawed them down to 1¾ in. by 5 in. I took the lengths and angles directly from the stringer, and sawed each segment so the grain would carry around the turns. I beveled the top edge of each section to get three surfaces of equal width, and routed a ¼-in. bevel on the bottom edges. All of the segments were predrilled to accept the dowels for the balusters. The joints were glued and screwed where they were easily accessible, and splined where it was harder to get at them. The connection between the horizontal rail at the top of the stairway and the first angled rail was made with dowels for the sake of simplicity.

The balusters are 1¼ in. square, and they are about 6 in. o.c., but because the sections of rail are different lengths and I wanted the balusters to be evenly spaced along each section, I had to vary the spacing a bit. I drilled ½-in. dia. holes 2 in. deep in both ends of each baluster to accept ½-in. dowels. Normally, I would mortise a baluster in, but the stairway's very shape helped make the rail so sturdy that the doweling, which is much simpler, was strong enough.

The balusters couldn't be cut to a single length to fit between the bannister and the stringer. I had to custom-cut each one to length, in the process making sure that I cut the correct angle on each one's top and bottom. To get the clean ⅛-in. bevels on all four edges, I hand-planed each baluster.

We had trouble deciding how to resolve the top end of the rail. The post ends at floor level with an oak and walnut burl cap sitting on top. At first we thought the rail would look good if it appeared to emerge out of the post. Well, once done, it didn't, so I brought out a chainsaw and did some remodeling. The result was to have a short 6-in. piece dip down at the same angle as the piece opposite where it starts to go down the stair. A much better solution. At the base of the stair, the rail simply ends with the last tread.   □

Tom Dahlke is currently building timber-frame houses in Canton, Conn. Photos by the author, except where noted.

Sculpted Stairway

Imagination and improvisation coax an organic shape from 42 plies of Honduras mahogany

by Jody Norskog

My brother Noel and I first met Chris Bradley and Steve Zoller in Santa Fe, N. Mex., in the summer of 1980. They were researching solar homes, looking for ways to improve a house they were designing and building, and generally checking out the Santa Fe scene. They were designing a house in Laguna Beach, Calif., and mentioned that they might have a stairway for us to build. We said sure, keep us in mind, and never expected to see them again.

Six months later we got a call from Laguna Beach; "Interested in doing that stairway?" Bradley and Zoller had been working on a design, but they weren't satisfied with the way it looked and wanted us to help.

Design concepts—Bradley and Zoller did not want a stairway framed up in the conventional fashion, and were unwilling to settle for a commercial spiral stair. They wanted a stairway that would be sculptural as well as functional—something that would be enjoyable to look at and use, but would not block the view from a large glass area.

Noel and I began debating design, materials and details. Originally we considered building the entire stairway out of steel. We later decided on wood after eliminating some other possibilities. After building models to study forms and detailing, we began exchanging ideas between Santa Fe and Laguna Beach.

The design we finally agreed on was far from conventional. We chose to support the treads and handrails with a single beam, which could be bent to form a compound curve. Imagine taking a tree trunk and bending it to the desired curve. And imagine notching the trunk and fastening a plank at each notch to make treads, and attaching a handrail on each side of the treads. This was the basic concept of the project. Our trunk would be a bent lamination.

Honduras mahogany was being used for the

Jody and Noel Norskog are partners in their firm, Norskog and Norskog. They build furniture and do architectural woodwork.

Looking as if it grew there, the Norskogs' mahogany and steel stairway winds its way gracefully to the second story. A mammoth undertaking, this stair took three months' work in their Santa Fe shop, and another six weeks at the Laguna Beach site.

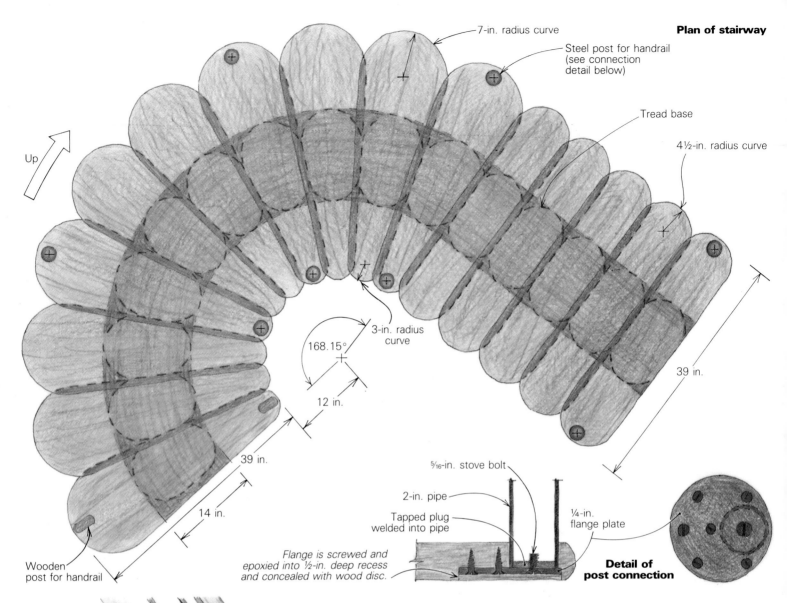

7-in. radius curve

Steel post for handrail
(see connection
detail below)

Tread base

4½-in. radius curve

Up

3-in. radius
curve

168.15°

12 in.

39 in.

39 in.

14 in.

Wooden
post for handrail

⁵⁄₁₆-in. stove bolt

2-in. pipe

Tapped plug
welded into pipe

¼-in.
flange plate

*Flange is screwed and
epoxied into ½-in. deep recess
and concealed with wood disc.*

**Detail of
post connection**

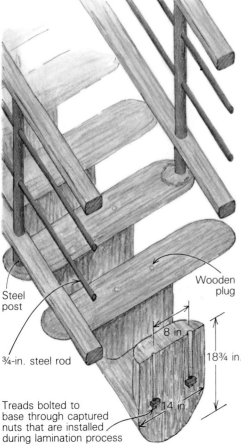

Steel
post

Wooden
plug

¾-in. steel rod

8 in.

18¾ in.

14 in.

Treads bolted to
base through captured
nuts that are installed
during lamination process

wood details and trim in the house, and it made sense to use it for the stairway as well, even though the wood had some bad qualities for the job at hand. While its crisp tissue is easy to carve and shape, it's not an easy wood to bend.

Once the design of the stairway had been reasonably defined, we worked with a structural engineer to determine what it would take to hold the thing up. After producing 30 pages of calculations, he was able to give us the information we needed—how to size the cross section of the trunk and how to design and size the hardware for connecting the trunk to the first and second floor.

The design and engineering process had stretched out over nine months before we got the go-ahead to build. Finding the wood for laminating the trunk turned out to be as hard as any other part of the project. No supplier in New Mexico could get what we needed. By doing some sample bending we determined the maximum thickness for each ply (or lamina) was ⁵⁄₁₆ in. Anything thicker was either too difficult to bend or too prone to tension failure— cracking on the convex side of the curve.

The problem was where to find pattern-grade Honduras mahogany boards ⁵⁄₁₆ in. thick by 15 in. wide by 18 ft. long. (The term *pattern grade* is peculiar to mahogany. It simply desig-

nates clear, stable stock, suitable for making foundry patterns.) Planing 4/4 stock down to ⁵⁄₁₆ in. would involve criminal waste, and resawing thicker, 18-ft. long boards was beyond the capability of our 14-in. bandsaw. We had to find a supplier who had lots of mahogany in the right dimensions. After much research and many telephone calls we located what we needed in a lumberyard in Long Beach, Calif. The lumberyard resawed 8/4 stock and surfaced the planks to our ⁵⁄₁₆-in. thickness. It was ironic that we bought the wood in southern California, had it shipped to Santa Fe, worked it, and then returned it to southern California.

Gluing up the trunk—The bending form for laminating the beam was an elegant piece in its own right (photo facing page). At either end there were plywood end plates with 2x4 platens for the actual working surfaces of the form, spanning between the plates. Essentially, it was like building a 2x4 stud wall to match the inside curve of the beam, and then leaning it on its side. Each 2x4 served as a clamping platen, and as a reference point for locating the position of the riser for each step. One person walked into the shop and said, "What are you guys building, an airplane?"

Few of the boards we got were wide or long enough, which meant that we had to butt sev-

Work on the staircase began with milling the stock and gluing up the 42 plies for the large lamination that would be sculpted to become the stringer or trunk for the treads. The bending form, though a simple jig made from 2x4s and plywood, had to be precisely built to produce the correct curve and twist. Because plastic-resin glue takes eight hours to cure, it took 42 days to glue up the whole lamination.

eral boards together to make up each layer of laminations. We were careful to stagger the joints from one layer to the next.

After getting the clamps, wood, and form ready, we did a dry-clamping run and felt we were ready for the first lamination. We mixed up the plastic-resin glue, applied it to the boards with paint rollers, put the wood on the form and clamped two plies together.

The next day we pulled the clamps off but the glue didn't hold. We tried it again with new boards. The same thing happened. At this point we were beginning to get frantic. Bad glue? Wrong kind of glue? Should we remill the wood? Then we discovered that plastic-resin glue requires a temperature above 70°F to cure. This was January at 7,000 ft. in northern New Mexico, where it gets cold. Our passive-solar shop maintains a comfortable temperature fairly easily—but not always 70°F. With our new kerosene heater we made several more tries, all the time refining our gluing procedure. At last we were able to get a bond that we felt good about.

Each tread was to be held in place by two countersunk bolts threaded into correspond-ing nuts embedded into the beam. The nuts were captured in the beam as it was being laminated. After the thirteenth layer, the first set of nuts was inserted. The 2x4 platens al-lowed us to position each nut accurately. We routed slots for the nuts and corresponding slots for bolt access. After each nut was set in the beam, the slots were taped over so they wouldn't fill with glue, and the laminating pro-cess continued. At the thirty-first layer, the second set of nuts was embedded into the beam. We used rectangular nuts made from ½-in. bar stock instead of standard hex nuts, to prevent them from turning.

Clamping time for Borden's plastic-resin glue is eight hours at 70°F. This meant that the whole lamination (42 plies in all) would grow at the rate of one ply per day. The pipe we were using for clamps could exert only enough pressure to hold down two layers of the mahogany over the curve. (We tried once to clamp three plies, but the force required was so great that the threads on the pipes be-gan to strip.) The first ply got the glue; the second we used as a caul to even out the clamping pressure. Each caul ply became the following day's gluing layer. Occasionally the cauls would fail in tension. There seemed to be no way of determining which board would fail except by testing it. Each day meant pull-ing the clamps off, sanding the entire surface, fitting boards for the next lamina, mixing and spreading glue, and clamping on the next ply. One lamination a day for 42 days.

Carving the trunk—Once the forty-second lamination was clamped down, it was time for a celebration. Even after it was notched and carved, the glued-up trunk weighed 2,000 lb., so it weighed considerably more at this point. We rented a 3,000-lb. capacity hoist for mov-ing the beam around. Used for pulling out automobile engines, it picked up the beam, bending form and all. Once the entire thing was in the air, it was easy to disassemble the form, piece by piece.

The location of the treads was marked by the 2x4s on the form. But after all the days of clamping, the jig was pretty well abused, and we couldn't trust its accuracy. With the assis-tance of the hoist, we placed the beam in up-right position. By drawing the plan view on

Once the lamination was complete, notches for the treads were roughed out with an electric chainsaw (top left). Then the tread-bearing surfaces were cut flat and level with a router attached to a plywood base (bottom left). With guide strips nailed to the trunk at the layout lines, it was easy to slide the router back and forth to mill the surfaces flat. To secure the stair to the first floor, a steel plate was bolted through to the framing. The bottom of the trunk was notched and relieved to receive a steel sleeve (above), which fits into the plate and is welded to it. The top of the stair is secured in similar fashion to the second floor.

the shop floor, we were able to transfer vertical riser locations up to the beam with plumb bobs and mark the horizontal cutting lines with levels.

With the rise and run determined, we used an electric chainsaw to rough out the notches for the treads to within a ½-in. of our layout lines (photo above left). To mill the horizontal surfaces of the notches flat, Noel made a long base for the router, attached guide boards to the trunk and surfaced the notches flat with a ⅞-in. straight bit, as shown in the photo below left. To give the massive trunk sculptural relief, we carved it to look like a tube, with the risers as intersecting tubes. The bulk of the carving was roughed out with the chainsaw, then cleaned up with various chisels and rasps. This exposed a pleasing lamination pattern, and led a number of people to ask how the risers were attached to the beam.

Making the treads was the only easy part of the project. Cut from 8/4 stock, they were laid out and arranged according to grain and color, and then planed down to a thickness of 1½ in. Their ends were rounded on the bandsaw and their edges routed to a half-round. The treads were then bolted to the trunk using the captured nuts that we had glued into the beam while it was being laminated.

To hold the stairway in place, our engineer called for a 10-in. square steel sleeve to be inserted into each end of the trunk. This was no problem at the top of the trunk because it was straight. But the curved bottom required a curved rectangular steel sleeve. We fabricated one (photo above) by heating four ¼-in. steel plates and bending them to the correct curvature before they were welded.

To avoid having to make separate bending forms for the handrails, and to get more accurate results, the Norskogs clamped L-shaped brackets to each tread at the location of the handrail (left), and then clamped the laminae to the brackets. Tubular-steel rails are bent to shape and then cut to length (top right). Then each rail is fitted between the posts and welded in place (above right) between the top and bottom rails of wood.

Handrail—The balustrade was designed to complement other handrails in the house—a wooden rail at top and bottom, with two intermediate ¾-in. steel-rod rails. The balusters we made from 2-in. steel pipe. Like the trunk, the wooden handrails are bent laminations. But instead of constructing a special bending form for gluing them up, we used the stairway itself as a form by making L-shaped brackets, and clamping one to each tread. Then we clamped the handrail plies to these brackets to get the right compound curvature for the rail (photo above left).

For laminating each handrail, we picked out an 8/4 board and ripped it into strips thin enough to bend to the required curve and twist. We kept the strips together in the same order in which they were cut, rolled glue on each strip, and clamped them back together. This procedure ensured that the original grain pattern would remain intact, and that the gluelines would not be highly visible.

The stairway changes from spiral to straight ten steps up. This transition made for some of the most difficult parts of the project. It causes problems for the handrail not only be-cause there is a transition from the curve to the straight, but also because there is a change in the angle of incline.

Looking at the stairway from the side view, the inside handrail has a convex transition and the outside a concave one. Instead of laminating these tricky compound shapes, we glued up stock large enough to carve solid sections as transitions between the curved and straight handrails.

Delivery and installation—This was also part of the job (a bigger part than we bar-gained for), so we rented a truck with an en-closed box. Loading the 2,000-lb. trunk was no problem with four men and a hoist on cast-ers. We were able to roll the trunk out the shop door right into the waiting truck.

Unloading was a different story. The house where the stairway was to be installed is on one of the world's steepest habitable hillsides. And there's no easy access, just a long walk almost straight up. We lost sleep for nights trying to figure out a suitable way to get the trunk off the truck, up the hill and into the house. When we arrived at the building site, we mustered all the men who were working on the house, and asked them to help haul the mahogany behemoth up the hill. We didn't tell them it weighed a full two thousand pounds. There was a lot of moaning and groaning, but we got it in place at last with nothing more sophisticated than grunt consciousness and brute force.

The installation went smoothly, but it took longer than we expected, as installations sometimes do. With the help of a welder, we set our mounting plates on the first and sec-ond floors (photo facing page, top right). The beam was lifted into place with another en-gine hoist. After positioning the trunk, the steel sleeves were attached to the ends of the beam. With the trunk in place, the steel sleeves were welded to the mounting plates. Final sanding on the trunk started, and treads were bolted down. Once the treads were down it was time to fit, mount, and shape the hand-rail (photos above right).

We returned to Santa Fe after six weeks in Laguna Beach. The stair was complete except for final sanding and finishing. The mahogany got six coats of Watco oil finish. □

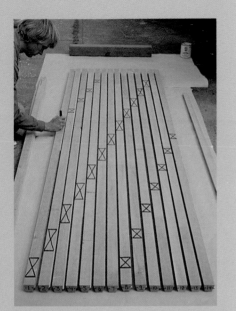

Coopered column. **The author marks the entry and exit mortises on the collated staves prior to precutting the mortises.**

Just a pinch. **To keep the staves intact during glue-up, White does not cut the mortises all the way through.**

Column glue-up. **Subassemblies of glued-up staves are clamped together so that their bottoms are in alignment.**

A Freestanding Spiral Stair

Using stock lumber and common shop tools to build an elegant circular staircase

by Steven M. White

As a woodworker, I have always liked curved staircases because they combine function with graceful, sculptural beauty. A spiral stair built around a central column is the simplest version of a curved staircase, but it's still a challenge to build. In this article I'll talk about a spiral stair that I've built for several clients and the straightforward techniques I've developed for fabricating its parts with common shop tools. For the twists involved in laying out a spiral stair, see the sidebar on p. 82.

Lightening the visual load—To my eye, most pole-supported spiral stairs are aesthetically flawed because they lack a graceful balustrade. Their handrails and balusters are often used as part of the structural support of the stair, and as a result they can look pretty clunky. To avoid this problem, my stair transfers the loads from the treads by way of cantilevered ribs that pass through a hollow column. This allows me to border the stair with a delicate row of balusters supporting a sinewy handrail (photo facing page).

Each tread/rib assembly resists three forces: the downward load of a person; twisting due to offset loading on the tread (the fact that a person would step near the front of a tread rather than directly over the rib); and horizontal, or back-and-forth, wiggle.

To resist the downward load, I use a strong, cantilevered rib that passes completely through the column and out the opposite side. Such a design transfers the load from the tread to purely vertical reactions on the column. It also allows the ribs to be revealed as tenons on the side opposite the treads. I put ½-in. chamfers on these tenons, emphasizing their sculptural qualities as they spiral from floor to floor.

For a 36-in. radius stair, I've found that a 2-in. by 6½-in. rib is adequately strong. Because the stresses on a cantilevered beam decrease with distance from the point of cantilever, I taper the ribs for the sake of appearance (top drawing, p. 81).

I also know that a 2-in. by 6½-in. rib will sufficiently resist twisting at its outer end, and that treads made from solid 2x lumber will be strong enough to withstand the offset load. To minimize horizontal wiggle, I rely primarily on the tread itself being drawn tightly against the column with lag screws. In addition, I run the front baluster of each tread through the tread and into the tread below. This interlocks all the treads and further reduces any wiggle.

For the stair featured in this article, I used vertical-grain Douglas fir throughout, which is a strong, attractive and economical wood. Incidentally, I bought unsurfaced stock for the ribs because it's a full 2 in. thick. Then I dressed it myself, netting ribs 1¹³/₁₆ in. thick.

Cylindrical column—The center column is round, with 24 mortises (two for each rib) in a precise spiral. To effect this, I built a 14-sided coopered cylinder in which the width of each stave is the same width as a rib.

First I ripped 9-ft. lengths of 2x6s into 1¹³/₁₆-in. wide staves, beveled inward at 12.86° on both edges. I got that number by dividing 14 into 360° and then dividing the result by 2. I fine-tuned the bevel by first running some scrap lumber and cross-cutting it into 14 sections to see how they fit together. To ensure correct alignment of all the staves during assembly, I cut a ¼-in. wide groove down the center of each beveled edge for plywood splines.

The next step was laying out and cutting the rib mortises. I placed all the staves side by side on a large, flat surface and numbered

Cantilevered treads. The author's spiral staircase rises on treads mounted atop ribs that extend through a hollow column. The treads wind their way from an enlarged first tread to a second-level landing that is borne primarily by ledgers affixed to the wall.

Doing the twist. Stretched across a bending form made of 2x4s on plywood bulkheads that duplicate the radius of the stair, thin laminations of Douglas fir are clamped and glued into a handrail blank.

Shaping the handrail. White makes several passes with a custom ogee bit to shape the handrail. The groove in the underside of the rail was cut with a 1-in. straight bit.

Handrail joint. Handrail sections are joined with a rail bolt. A hole in the underside of the upper rail allows access to tighten a nut onto the bolt's machine-threaded end.

Rail return. At the landing, the handrail engages the wall by way of a curved section cut from a solid piece of stock. Its bolt hole is plugged with a dowel.

them to avoid any confusion later. On each stave I marked a 6½-in. high mortise for the front side of one rib and a 3-in. high mortise for the protruding tenon of another rib. These mortise locations were laid out sequentially, at 8.156-in. intervals, corresponding to the rise of the stair (top photo, p. 78). I then cross-cut each mortise with a power miter saw, being sure to leave a little wood uncut on the backside to keep the stave in one piece (middle photo, p. 78).

Now I was ready to assemble the column. I began by gluing up trios of staves, using pipe clamps to hold them together. I left the alignment splines and glue out of the mortise regions to ease the cleanout still to come. Then I glued up the four sub-assemblies using twisted ropes as clamps (bottom photo, p. 78). Fourteen staves at 1¹³⁄₁₆-in. width yielded a column diameter of about 8 in.

Once the glue had dried, it was a simple matter to split out the mortise waste with a chisel. The resulting holes, however, were narrower on the inside than on the outside because of the stave being beveled, so I pared out the sides of the mortises with a chisel. I used a test rib to check for a snug, smooth fit through each mortise.

To round the column, I planed down the 14 corners by hand to create a 28-sided cylinder. Then I used my plane and sandpaper to work the assembly into a smooth cylinder.

Treads and ribs—I began the tread part of the project with a cardboard template. Each tread is ¹⁄₁₄ of a circle, or 25.71° of arc, and my template started as a simple pie slice at this angle with a 36-in. radius. To this I added 1¾ in. of width to both the front and back edges, to give a generous nosing and to widen the narrow end of the tread. The narrow end of each tread is curved to match the column's face. The curved corners flare into a straight cut that allows the tread to be let into dados cut into the column (large photo, p. 83).

I glued up pairs of 2x12s to get the wide stock required for wedge-shaped treads. Then I sawed out the treads on the diagonal, making sure that the front edge of each tread was parallel to the grain. A trammel-point jig on my bandsaw helped me to cut smooth outside arcs on the treads.

All the treads are identical except for the starting tread and the landing tread (drawing right). Traditional staircases usually feature a prominent starting step, so I widened the first tread toward the newel post. I also widened the top landing to meet the door opening. The landing is flush with the second-story subfloor so that it can be carpeted to match the upstairs-hall floor.

I used a ¾-in. roundover bit to soften the edges of the ribs and treads. After a good sanding, I glued and screwed each tread to its rib. Each rib is under the centerline of its tread.

The balustrade—I make spiral handrails by gluing up thin laminations over a bending form (top photo, facing page). This is a full-

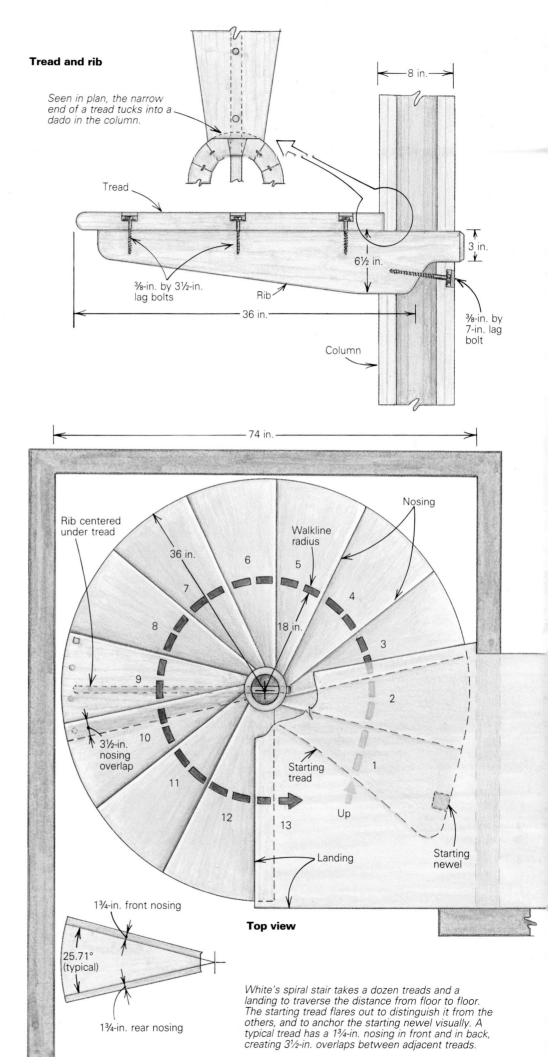

Tread and rib

Seen in plan, the narrow end of a tread tucks into a dado in the column.

Tread

⅜-in. by 3½-in. lag bolts

Rib

Column

8 in.

3 in.

6½ in.

36 in.

⅜-in. by 7-in. lag bolt

74 in.

Nosing

Rib centered under tread

36 in.

Walkline radius

18 in.

3½-in. nosing overlap

Starting tread

Up

Landing

Starting newel

1¾-in. front nosing

25.71° (typical)

1¾-in. rear nosing

Top view

White's spiral stair takes a dozen treads and a landing to traverse the distance from floor to floor. The starting tread flares out to distinguish it from the others, and to anchor the starting newel visually. A typical tread has a 1¾-in. nosing in front and in back, creating 3½-in. overlaps between adjacent treads.

Spiral-stair layout

For a spiral stair of any given diameter and floor-to-floor rise, there is an optimum unit rise and number of treads per revolution. You find these two numbers by applying the standard rise-to-run formula to the stair at its walkline.

The rise-to-run formula is: two times the rise plus the run should total 24 in. to 25 in. Also, the maximum comfortable incline for a spiral stair is considered to be 45°, at which point the rise equals the run. In our formula, a 45° stairway would result if the rise and run were both 8 in. to 8⅓ in.

The walkline is the line followed by a person using the stair, and for treads less than 36 in. wide this is generally taken as the centerline of the stair. The length of the walkline provides the basic measure from which the length of run is calculated. This stair has an outside diameter of 72 in., or a radius of 36 in. Its walkline radius is therefore 18 in., and the walkline length in one full revolution is the circumference of a circle with an 18-in. radius. The circumference of a circle is found with the formula 2pi x radius. For this stair, the walkline circumference is therefore 113 in. The minimum desirable run of a tread is about 8 in., so I divided 113 in. by 8 in. to get the number of treads needed per revolution for the stair. The answer is 14.1, which I rounded to 14 treads per revolution.

This staircase has a total rise of 106 in. The fact that the run was about 8 in. dictated a rise of about 8 in. Because 106 ÷ 8 = 13.25 rises, I rounded down to 13 rises, resulting in a unit rise of 8.156 in. There were to be 13 rises, so the stair would spiral less than one revolution. It would have 12 treads, with the 13th step being the landing.

One final layout concern should not be overlooked. I have seen designers specify a 12-tread-per-revolution stair with a 90° arc (quarter-circle) landing at the top. This is generally not desirable because of headroom considerations. Imagine descending such a stair. As you reach the front edge of the ninth tread down, you are faced with the back edge of the landing above you. As you step off the ninth tread, will your head clear the landing? It will—but only if you're short. The accepted figure for headroom on a stair is 80 in. In order for nine steps to drop you 80 in., the unit rise would have to be about 9 in., assuming the landing had very little thickness to further encroach on headroom. Most codes allow a 9-in. rise for a spiral stair, but as I explained above, such a stair would be very steep. The best solution to this problem is to cut back the landing to a 60° arc.

An excellent book on spiral-stair layout is *Designing Staircases*, by Willibald Mannes, (Van Nostrand Reinhold, out of print). In it, Mannes presents diagrams summarizing the optimum layout for stairs of varying diameters and heights. —S. W.

size cylinder representing the diameter and height of the staircase, and is made of spaced 2x4s screwed to plywood bulkheads laid on sawhorses. Because I was building an outer handrail, I built the cylinder to the exact radius of the inside of the handrailing, in this case 33⅛ in. I built the rail in two halves, so the form needed to be just half a cylinder, and only half the stair height—about 5 ft.

Calculating the angle at which the rail spirals up the cylinder is a straightforward rise/run calculation. In this case the unit rise was 8.156 in. and the unit run was 1/14th the circumference of a 33⅛-in. radius circle, or 1/14 (2pi x r), which equals 14.87 in. Knowing the rise and run, the angle of ascent is tan⁻¹ (rise/run) = 28.75°. With this information I plotted reference points on the 2x4s to guide the placement of the handrail laminations.

Starting with 10-ft. long Douglas fir stock 2⅜ in. thick, I ripped nine laminations, each ¼ in. thick, for a 2¼-in. wide rail. Keeping the laminations in order makes for less-visible glue lines later. I use a ripping blade on my table saw and glue the sawn laminations unplaned.

I have always managed to glue up all the plies of a spiral rail at once, but I have to work fast. I lay out all my pipe clamps and C-clamps near the cylinder, set all the laminations side by side in the right order on a flat surface with the center marked on each piece, and check that I have plenty of glue.

One time, incredibly, I ran out of glue partway into the spreading. Luckily, a friend was helping, and he spread what glue we had while I raced to the nearest hardware store, where I stood in line, twitching with adrenalin flashes as I pondered the uses of a partially laminated handrail. But I only lost about five minutes, and the glue-up was saved.

For handrail laminations, I use a toothbrush to apply aliphatic resin glue. I then stack the laminations in order and carry them to the cylinder. I first clamp the center, then work outwards, clamping the rail to the 2x4s and aligning their edges with the reference points. Bending 2¼-inches worth of laminations can take a lot of force, and pipe clamps are handy for the initial bending. Then I check for gaps between plies and clamp wherever I find one—any unglued gaps can later split the whole rail open. The 40-or-so clamps I have are barely enough for a 10-ft. bend. Once the rail is aligned and clamped to my satisfaction, I let it cure for 24 hours. With this staircase, I repeated the operation for the other length of rail.

I clamp the cured handrail in my bench vise and square up the rail cross section with rasps, planes and scrapers. This operation is easy enough, but the next operation—shaping the rail—is not. Although I have a custom-made ogee router bit (made by Oakland Carbide, 1232 51st Ave., Oakland, Calif. 94601; 415-532-7669) to shape the rail with a minimum of passes, the rail has no flat surface to guide the router. I compensate for this by warping the plastic router base with shims to conform to the twist of the rail (bottom left photo, p. 80). But the operation requires a steady hand. After

routing, I use sandpaper and plenty of elbow grease to finish the handrail sections. Last, the upper length of handrail needed a wall-return piece at the landing (bottom right photo, p. 80), which I attached with a standard rail bolt.

The balusters on this stair are 1-in. square spindles. On any stair or landing, I try to keep the space between balusters to no more than 4½ in. both for the safety of any children using the stair and for looks. In this case I used three balusters per tread. I machined round tenons at the base of each baluster to fit into holes drilled into the treads, but left the tops long to be cut on site.

This stair has three newel posts: a starting newel and two landing newels. I made them from 4x4s, bandsawing a four-sided taper in the midsection (photo, p. 78). A commercially made ball top caps each newel. One of the landing newels sits atop the center column and has a tenon at its base to fit into the column.

Pulling it all together—I began the installation with the second-floor landing. Ledgers screwed to adjacent walls provided support for three of its four corners. The fourth corner bore on a dado in the stair column. This intersection was further strengthened with a lag bolt.

Once I had the landing in place, I temporarily installed the column. Making sure that the column was plumb, I traced around it at the floor. Then I removed the column and screwed a wooden disc the diameter of the inside of the column to the floor. I slipped the column over over this disc and tethered the top of the column to the landing with a rope. This allowed me to rotate the column as I inserted the treads. A rotating column isn't necessary if there is good access to all sides of the stair.

Installing the treads is the fun part of the job (large photo, facing page). I start at the bottom, pushing each one home and then drawing it tight to the column with a ⅜-in. by 7-in. lag bolt. Once all the treads are in place, I permanently lag-bolt the column to the landing and plug the recessed bolt holes with dowels.

Balustrade assembly began with the newel posts. Once they were in place, I was ready to join the two lengths of spiral handrail. To determine their finished length, I laid them in position atop the treads and marked them where they overlapped. Then I square-cut the two abutting ends and joined the pieces with glue and a rail bolt (bottom middle photo, p. 80). Next, I bevel-cut the tops of several balusters to 32 in. These balusters were inserted through the front hole of several treads and temporarily braced plumb. This would establish the top of the handrail at 34 in. above the nosings of the treads. Setting the spiral rail on top of the balusters, I screwed the downstairs end of the handrail to the starting newel. This intersection is further strengthened with a ½-in. dowel. Upstairs, the handrail engages the wall with a return that is Molly-bolted to the wall. Toenailing the balusters into the rail fixed it in position, and I could then fill in the rest of the balusters. I measured and cut each one individually to ensure a tight, plumb fit.

Rail-to-newel. At the starting newel, the handrail is affixed with a ½-in. dowel and a 2-in. galvanized drywall screw.

The fun part. Once the column has been slipped over its base and tethered to the landing, White can insert the numbered treads into their mortises, and the stair quickly takes shape. Note the dados above the mortises at the top of the column. The dados accept the ends of the treads, making a clean joint and reducing wiggle.

Pass-through balusters. The leading baluster on each tread passes through a mortise to engage the tread below it. White used a 1-in. chisel to square the mortises.

The balusters that extend through the front of each tread and into the tread below were fixed in place by dowels driven horizontally into each tread. They help link the entire assembly into a single unit, reducing the tendency of individual treads to wiggle as they are stepped on. Although it looks like it would be impossible to install these balusters with the handrail in place, there was no problem. They are flexible enough to bend out of the plane of the handrail as they are driven home with a hammer and a wooden block (photo above).

To stiffen the balustrade at the bottom of the stair, I drove a ⅜-in. steel pin through the rail and into a stud in the wall. Fortuitously, a stud was located adjacent to the rail's closest proximity to the wall.

I finished up by fairing the joint between the handrail sections to make a smooth transition. Then I applied three coats of Daly's Floor-Fin (a urethane and tung-oil finish) to the entire stair (Daly's Inc., 3525 Stone Way N., Seattle, Wash. 98103; 800-521-0714). I like the way this finish looks, and it's easy to apply—it's a brush-on and wipe-off finish. It's also formulated to hold up to the rigors of foot traffic. □

Steven M. White is a designer, carpenter and woodworker specializing in staircases. He lives in Berkeley, Calif. Photos by Charles Miller.

Concrete Spiral Staircase

A massive stair made by casting treads in a precision mold and bending thin-wall tubing

by Dennis Allen

The stair's concrete treads are threaded on a steel column and locked in place by steel balusters. Each of the 300-lb. steps was cast individually in a mold lined with plastic laminate.

This circular staircase was inspired by the stone stairways built in Europe during the Middle Ages. Designer Paul Tuttle wanted to create the sense of timeless solidity that massive stone steps evoke, and the two-story greenhouse in the Douglas residence overlooking Santa Barbara, Calif., gave him his chance. The room needed both a stairway and a sculptural focus, so Tuttle captured the medieval aura with the 10-ft. high, 7-ft. dia. concrete spiral stair shown in the photo at left.

Even before I had seen the design, Tuttle asked me if I would be interested in building this staircase. My first impulse was to say no. A poured-concrete spiral stairway seemed impossibly difficult. But once I saw his drawings, I got excited by the challenges its construction presented. Along with two of my associates, I agreed to build it. But I still wasn't really sure I could.

Layout—The first thing we had to do was to determine the number of steps, the rise of each step and where the beginning and end of the spiral would fall. We eventually decided on a landing and 15 steps of 7½-in. rise. Each one is a 25° segment of a circle.

Next we made a full-scale mechanical drawing of one of the steps, including sleeves for holding the balusters, a cutout for toe space, indentations for carpeting and a sleeve for the center column. This drawing (facing page, left) proved to be an indispensable reference in all stages of the project.

Making the concrete form—After we figured out what one of our concrete treads had to look like, we had to build a form that would duplicate it 15 times. The form had to be durable, yet flexible enough to be taken apart after each pour and reassembled for the next one. Carpenter John Bunyan and I thought about our options and eventually came up with the one shown on the facing page, right.

The form's main components were the base, the sides, the pivot end and the outer end. The base was a 3-ft. by 4-ft. piece of ½-in. plywood that we covered with plastic laminate (for more about working with plastic laminates, see *FHB* #9, pp. 39-41). We made the

Mechanical drawing of a step

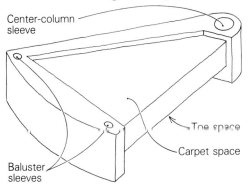

Center-column sleeve

Toe space

Carpet space

Baluster sleeves

Steelwork

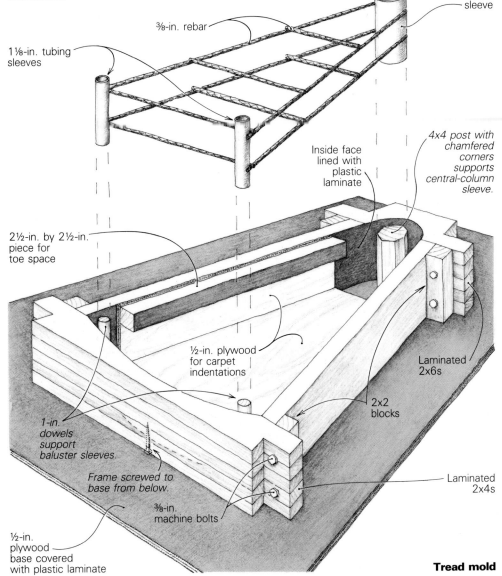

3/8-in. rebar

4-in. pipe sleeve

1 1/8-in. tubing sleeves

Inside face lined with plastic laminate

4x4 post with chamfered corners supports central-column sleeve.

2 1/2-in. by 2 1/2-in. piece for toe space

1/2-in. plywood for carpet indentations

Laminated 2x6s

2x2 blocks

1-in. dowels support baluster sleeves.

Frame screwed to base from below.

3/8-in. machine bolts

Laminated 2x4s

1/2-in. plywood base covered with plastic laminate

Tread mold

form's sides out of fir 2x8s, with 2x2 blocks screwed and glued to their ends. We made the tight-radius pivot end from five 2x6 blocks laminated together with yellow glue. We cut out its 4-in. radius arc on the bandsaw. The outer end was made up of five 2x4s. It had a 21-in. arc of 3½-ft. radius cut out of it. We let the tails run long on the end pieces so that they could be bolted to the end blocks on the sides of the form.

Once we had the two sides and the tight-radius end piece bolted together, we lined its inside face with plastic laminate. This part of the form stayed together as a unit throughout the casting process. We lined the outer end, and then bolted it to the side pieces. We put the entire assembly on the base, and screwed the two together with 26 1½-in. screws.

Next we screwed plywood pieces to the bottom and lead edge of the form to create an indentation for carpeting. The shape of these pieces had to be carefully worked out so that the carpeting would flow from one step to the next without any offset. We planned for the bottom of the form to be the mold for the top of each step, so we could place the largest of the ½-in. let-in pieces for carpeting on the bottom of the form. Pouring the steps upside down also let us trowel the bottom of each tread—the largest and most visible expanse of concrete on each step. Finally, a 2½-in. square piece of wood 28 in. long was covered with laminate and screwed to the top edge of the form. This would form the indented toe space on the bottom lead edge of each step.

The last parts of the form were the registration pins for the steelwork—two 1⅛-in. tubing sleeves to support the rebar, and the 4-in. pipe sleeve that would fit over the central column of the stairway. We screwed two 7½-in. lengths of 1-in. wood dowel to the base near the corners at the wide end of the form (drawing, above right). At the other end, we attached a 7½-in. piece of 4x4 with chamfered corners. These pegs, carpet inserts and toe board were screwed in from the outside of the form so they could be easily released when we stripped the form from the cured concrete.

Structural steel work and balusters—We decided to cast a short length of 3-in. pipe into the slab-floor footing to act as an anchor for the 3½-in. central-column pipe. Each step would have cast within it a sleeve of 4-in. pipe. These sleeves would slip over the center

column and become an integral part of the rebar assembly in each concrete tread. The rebar grid would also include the two 1⅛-in. tubing sleeves into which the railing balusters would slide and be secured, as shown in the drawing above..

The sleeves in each step are attached to one another by a matrix of ⅜-in. rebar. Fifteen grid assemblies were required, one for each stair, and each one had to match all the others exactly in order for the balusters and central column to fit properly.

To achieve this kind of accuracy, our steel expert, Dean Upton, welded each assembly on a jig. In a ⅜-in. steel plate, Upton drilled holes at the sleeve centers and tapped them for ½-in. bolts. Then he tack-welded posts to the plate. These posts had been turned to fit the inside diameter of the sleeves, and were bored to accept the ½-in. bolts. The sleeves for each tread were cut to length and squared. Then Upton bolted them to the jig and welded the rebar in place.

The posts that support the handrail are 1-in. OD steel tubing with a .083-in. wall. The sleeves into which they fit are 1⅛-in. OD tubing with a .049 in. wall, allowing a clearance of .027 in. That seemed a bit loose to us, but

proved to be a necessary margin during the final stair assembly.

Upton silver-soldered a ¼-in. ring (from the sleeve material) to each baluster to make sure they ended up at the right height. Their tops were cut at the angle of the stairway, and as they were installed, the balusters were rotated to match the direction of the handrail. Once the treads were in place, we plugged the underside of the baluster holes with Bondo (a brand of auto-body putty).

Picking out the pipes—Material selection required some careful sleuthing. Our basic plan called for several pipe sizes, each to fit snugly over the next. The central column is 3½-in. pipe, schedule 40. The sleeves for the steps are 4-in. pipe, also schedule 40. Pipe goes by the inside diameter (ID) while tubing goes by the outside diameter (OD). With pipe, oddly enough, the OD is constant while the ID changes with the schedule or wall thickness. A 3½-in. pipe (schedule 40) is actually 4-in. OD by 3.548-in. ID, with a wall thickness of .226 in. So depending on the schedule, we had a choice of clearances between sleeve and column. With 3½-in. pipe and a 4-in. sleeve, there is a theoretical clearance of .026 in. But

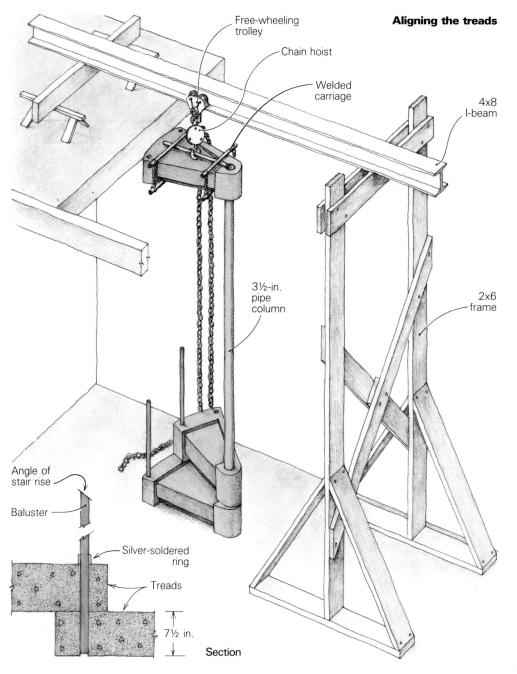

Free-wheeling trolley

Chain hoist

Welded carriage

4x8 I-beam

3½-in. pipe column

2x6 frame

Angle of stair rise

Baluster

Silver-soldered ring

Treads

7½ in.

Section

Assembly—Once the 15 treads were cast and carted to the site, our next hurdle presented itself—slipping the 300-lb. steps onto the steel column. Obviously we needed some type of device to lift the treads. It would have to be sturdy enough to carry the heavy loads, yet also adjustable so we could fine-tune the position of each tread over the center shaft.

Our solution was the homemade chain hoist shown in the drawing at left. It consisted of a 12-ft. 4x8 I-beam and a chain hoist mounted to a freewheeling trolley. We centered the beam over the column, and held it up by the stair landing on one side and a sturdy framework of 2x6s on the open end.

We fabricated a special metal carriage and harness to carry the treads as close as possible to their center of gravity. Each tread was then lifted above the 10-ft. high column, and its sleeve was centered over the shaft. Then the tread was slowly lowered into position. Once a step was in place, we would brace its outer end with a 2x4 and then one of us would tap a baluster through the aligned sleeves to secure the new tread to the one below it.

Disaster nearly befell us midway in the assembly. As we rolled the tenth tread along the I-beam, a lurch in our movements caused a sudden shift in the tread's center of gravity. Instantly the harness slipped off the carriage and the step plummeted, bouncing against several of the steps already in position and demolishing the bottom picket on its way to the floor. We were stunned. Fortunately nobody had been standing in the way of the tread when it fell. We surveyed the damage and it appeared enormous. Chunks of concrete were knocked off the treads in half a dozen places. We were so badly shaken that we packed up and went home for the day, believing the project ruined.

The next day we reassessed the damage and concluded that it wasn't as severe as it had seemed. We decided to patch the damaged edges and corners with Bondo. In some places we had to build up numerous layers of the stuff, but it worked far better than we had dared to hope. Because the Bondo was a different color from the pristine white concrete, we knew we'd have to paint the final product.

At the landing—The top step had a different shape from the others because it needed to flow into the cantilevered landing. To link the stairway to the landing, we built a triangular rebar grid to lock the top of the central steel column rigidly to the 4x12 landing girders. We welded sleeves for the balusters and the central column to this grid. The grid in turn was welded to a 4-in. by 30-in. by ⅜-in. steel plate, which was bolted to the landing framing. Then we erected a form around the grid with supports down to the floor. We were able to use several curved components from our breakdown form, but most of the pieces were new and had to be covered with plastic laminate.

We poured this last step with the same two mixes and care that we used with all the other treads, and when we took off the forms it flowed perfectly into the landing.

this clearance did not prevail on all of the pieces so some sleeves had to be turned down on the lathe.

Concrete technology—The concrete mix was critical for two reasons: weight and finish texture. In order to decrease the weight of each tread from more than 400 lb. using concrete with standard aggregate to about 300 lb., we used ½-in. Rocklite (The Lightweight Processing Co., 715 N. Central Ave., Suite 321, Glendale, Calif. 91203), a lightweight aggregate. But we also wanted a dense, pure white finish on each tread. This led to our using two different batches for each one. The outer inch or so is made up of 1 part white portland cement, 2½ parts 60-grit silicon sand and 2½ parts Cal-White marble sand (used mainly for swimming pools), made by Partin Limestone Products Inc. (PO Box 637, Lucerne Valley, Calif. 92356). Once we got this outer layer of white concrete in place, we filled the core of each tread with the lightweight mix.

We carefully measured all the ingredients because slump was important—too much slump would cause the two mixes to flow together in the form. The lightweight mix for the core was 1 part cement, 2½ parts sand and 2 parts aggregate. To speed setting time, we added a little calcium chloride to each batch.

Originally we'd hoped to pour two steps per day, but found that producing one a day was quite an accomplishment. Placing the mixes in the form required two of us—one to tamp the outer mix and the other to keep the core mix from migrating to the edge of the form. We placed the concrete in layers, agitating it thoroughly after each layer to eliminate voids. Between each pour we cleaned the form, coated all dowels and wood insets with floor wax and sprayed the plastic laminate with silicone.

Surprise and delight filled us when we stripped the form from the first tread. The result was magnificent, but not what we'd expected. The plastic laminate made the surface smooth as glass, and a swarm of tiny, irregular air pockets made it look something like travertine. We were elated with this first success.

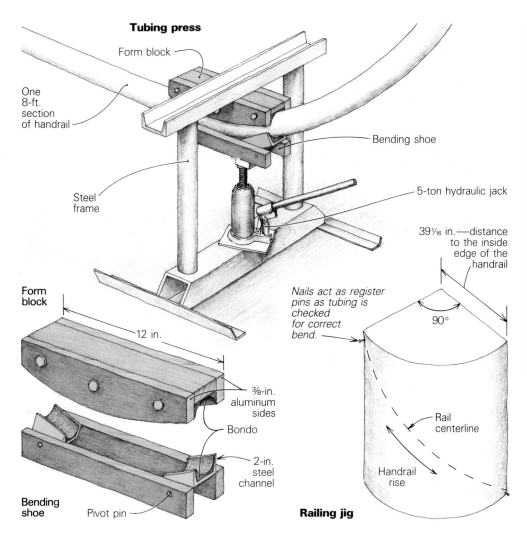

Tubing press

Form block

One 8-ft. section of handrail

Steel frame

Bending shoe

5-ton hydraulic jack

Form block

12 in.

⅜-in. aluminum sides

Bondo

2-in. steel channel

Bending shoe

Pivot pin

Nails act as register pins as tubing is checked for correct bend.

39 1/16 in.—distance to the inside edge of the handrail

90°

Rail centerline

Handrail rise

Railing jig

Dean Upton torch-welds the handrail to a steel baluster. Although it appears continuous, the railing is composed of short segments of steel tubing that were bent on a homemade press, and then assembled on site to create the necessary helical shape.

Handrail—Probably the most challenging part of this project was bending a 1⅞-in. dia. thin-wall tube (.063 in.) into a helical handrail. We chose this size tubing because there are stock fittings for 1½-in. pipe that fit closely enough to be used with the tubing (1.875-in. OD vs. 1.900-in. OD). At the top where the stairs meet the landing, we needed a tight return bend to blend the rising stair rail into the horizontal landing rail. We made this transition with two wide-radius elbows and a little cutting and fitting. We used another stock fitting—the half-sphere cap—to finish the bottom end of the handrail, and we used floor flanges to attach the landing rails to the wall.

The radius of the stair circle was 42 in., but the radius of the line of balusters was 40 in. The inside radius of the handrail was 40 in. less 15/16 in. (half the diameter of the tubing) or 39 1/16 in. Taking his cue from a tubing bender, Upton designed a press that used a hydraulic jack to generate the bending force needed to arc the straight lengths of tubing. He used an oak form block with a radius of 38⅞ in., a little tighter than the required radius to allow for some springback. As it turned out, the springback was almost nil.

The forming tool we tried out first had two spools about 12 in. apart. It bent the tubing, but it also left slight dimples at each point of contact between spool and tube. A handrail with a dimple every 6 in. was totally unacceptable (it looked like a segmented worm), so Upton made a pair of bending shoes out of

channel steel and Bondo as an alternative. They needed periodic greasing to allow the tubing to slip through as it was bent. This jig, shown in the drawing above left, worked fine. A 5-ton jack supplied the pressure.

Upton first tried to form the helix as the tube was being bent by rotating the tube a little at each bend. But it was difficult to keep track of the rotation. We could calculate how much rotation was required, but to control it was tough in a small shop. Even though it wasn't the right shape for our railing, the sculpture resulting from the first try could be mounted on a stone block and placed in front of a library.

We learned two things from this attempt. One, the press could put a wrinkle-free radius in our tubing, and two, trying to form both the radius and the helix into the full-length railing was too ambitious. Instead, Upton cut the 24-ft. tube into three 8-ft. pieces. Then he made a plywood jig that had a radius of 39 1/16 in. (drawing, above right). This jig represented about ⅓ of a turn of the staircase, and was tall enough to allow the rise of the handrail to be marked diagonally on it. As he shaped each 8-ft. section, Upton checked its bend against the jig. This worked well, and the three pieces closely approximated the required helix plus the radius.

To make final adjustments in the helical twist, each 8-ft. section was cut into three equal pieces. After tack-welding the first section to the bottom balusters, Upton rotated

the second section slightly to create the helix. Section two was then tack-welded in place, and the third piece rotated slightly more than the second and so on until all nine parts were tack-welded in place. Each piece was aligned with its neighbor by using a short offcut of 1¾ in. tube as a dowel. Before he welded the balusters to the railing (photo above), Upton torch-welded the whole unit into one continuous piece. Then all the welds were ground down, and any little pits were filled with Bondo and sanded smooth. The finished rail is painted brick red, and appears to flow as one piece from top to bottom.

Our final job was whitewashing the treads. We wanted to preserve the texture of the concrete and to have it not look painted, so we experimented with several finishes. We finally settled on white latex paint mixed with a small amount of white portland cement. This gave the surface a little roughness to the eye, but did not destroy the glass-smooth texture to the touch. One coat completely covered the grey-green Bondo, and we were done.

The project took six weeks of concentrated effort, and it kept our attention with a series of snags and surprises. But everybody is happy with the way it turned out. The stairway cost almost $10,000—a lot for one flight of stairs, but not for a sculpture that anchors a special room.   □

Dennis Allen is a general contractor living in Santa Barbara, Calif.

Circular-Stair Retrofit

Wrapping concrete with wood called for diverse techniques and a common bond

by John Alexander

Construction of the Arkansas Governor's Mansion in Little Rock began in 1947 and continued on into 1950. But money ran out before the cantilevered circular stair in the entry foyer could be completed. The poured-concrete stair was simply carpeted over and fitted with a plain metal handrail and metal balusters, and for almost 40 years, that's the way it stood. A while back, my architectural-woodworking company was asked to look at the stair and determine whether it was feasible to build an elegant wood stair over it.

The stair begins in the basement of the mansion and winds in three separate flights through the foyer to the second floor. Our proposal called for the removal of the old carpeting and metal railings and for the installation of wood treads, risers and decorative stringers over the existing stair. Also, the old skirtboard along the wall would be removed and a new one notched into place. The metal railing would be replaced by an over-the-post balustrade. Except for the balusters and newel posts, we'd manufacture all stair parts and trim in our shop.

Measuring up—Because we were building off an existing stair, we had a lot of preliminary measuring to do. The first step was to establish a point of reference. We started by propping a 2x4 over the stairwell, resting one end on a ladder and the other end on the balcony railing. Then we dropped a plumb bob from the 2x4 down to the basement floor and measured and jockeyed the plumb bob until we located the center of the stairwell. After marking the floor at the center point, we pulled a wire down from the 2x4 to the mark and fastened it to a wood crosspiece weighted down with concrete blocks. A turnbuckle served to stretch the wire taut. Now we could measure the inside and the outside diameters of the staircase at every riser, using the wire as a datum line.

For the vertical measurements we rented a builder's transit (for more on levels and transits, see *FHB* #37, pp. 39-45). We established our first base at the basement floor, then shot and recorded the elevation of each step. Once we shot as many of the treads as we could from the basement floor, we moved the transit up to the first landing at the top of the basement flight. Then we reshot the previous tread, thus establishing a new base. In this way we worked our way up to the top of the staircase, landing by landing. We would use all these figures later to help us build the bending forms for the glue-laminated stair parts and to lay out the stringers and skirtboards.

Once we were through measuring, we duplicated in cardboard the six sweeping treads of the main flight located in the foyer. Because each of the six steps was of a different radius and each would require a curved riser, this was the most complicated flight in the staircase (photo facing page). We also made a cardboard pattern of the reverse curve at the top balcony.

Drawings—It took a full week to digest the data and to translate it into a drawing of the stair. I first drew the stair we wanted to end up with, then superimposed the existing stair in red over it. That way we could see the problem areas. The goal was to adjust the layout so the risers in each flight would be identical, and all but the six curved treads in the main flight would be the same except for the outboard ends. These were left long and would be scribed to the skirtboard and cut to fit later. As any good stairbuilder will tell you, variations in stair size can cause tripping, and except for the bottoms or tops of flights, most building departments won't allow treads or risers to vary by more than ¼ in.

Most of the problems we encountered were in the basement. There was a 5-in. bulge in the wall along the outside radius of the stair, several of the treads were too high (up to ¾ in. relative to the mean) and the whole flight of 15 steps was skewed so that the risers did not point to a common center as they should have.

But I was most interested in the first six steps of the main flight. The original concrete steps were a little out of kilter, and I hoped to find a mathematical relationship that would allow us to lay out the curved treads and risers accurately so they would flow around the existing concrete with a minimum of concrete removal. My brother and I solved the problem with an Apple computer and with a program in BASIC that my brother wrote. The computer served as a fast, flexible drawing board, which allowed us to draw and erase various curves and to compare them to the arcs of our cardboard templates. We could display one tread on the screen or all of them at once. The computer helped us to establish a logical layout that we would later render in full scale with trammel points and a tape measure.

Bending the parts—The new stair combined several different kinds of wood. We settled on quartersawn white oak for the treads and walnut for the railing, all of which would receive a natural finish. For the paint-grade risers, stringers, skirtboards and trim, we chose poplar. We ordered paint-grade birch balusters and newels from Morgan Products, Ltd. (P. O. Box 2446, Oshkosh, Wis. 54903).

We built two forms for the curved parts—one for the stringers, railings and trim on the inside edge of the stair, and another for the skirtboards and trim to be installed along the wall. Based on our earlier measurements, the average inside diameter of the existing stair was 41½ in. The new ¾-in. thick stringers would reduce that to 40¾ in. The inside diameter of the handrail would be ½ in. smaller than that, so we made the inside form at the smaller diameter and simply added ½-in. spacers when it came time to bend the stringers and trim pieces. The average outside diameter was 94½ in., so we built the outside form with a 93¾ in. dia. to compensate for the ¾-in. thickness of the skirtboards and trim.

We determined that the taper of the treads should be 9°, and the run of each tread should be 6⅜ in. on the inside end and 14²³⁄₃₂ in. on the outside end. We drew the two circumferences on the floor of our wood-storage building, marked the riser locations with dividers and screwed an upright 2x6 to a 2x cleat anchored to the floor at each mark. We screwed a block to each upright at the proper elevation so that the pieces we were gluing up would rest on the blocks. To accommodate the longest flight, we built the forms 10 ft. high and reinforced them with two ¾-in. plywood semicircles spaced about 3 ft. apart.

Everything but the handrail was laminated out of three ¼-in. thick layers of poplar, arranged so the best face would be visible. For each curved piece, we first spread white glue on the plies with a paint roller (white glue gave us a longer clamping time than yellow glue). At once, we clamped all three plies to the 2x6s with bar clamps. Between the bar clamps we used plenty of C-clamps; 2x2 cauls served to distribute the pressure.

Before gluing up the skirtboard and corresponding trim for the basement flight, we added blocks to our form in order to duplicate the 5-in. bulge in the basement wall. We also adjusted the forms to make the pieces for the second-floor balcony.

The handrail was made of 10 plies of ¼-in. walnut, each slightly under ¼ in. thick. Here we used plastic resin glue instead of white glue so the glue lines wouldn't show.

Planing, sawing and routing—Once pieces were laminated, we moved them over to the shop and squared up one edge with a 3⅝-in. power plane. The plane was fitted with wooden fences, which were sculpted to match the various curves. Then we ripped the skirtboards, stringers and trim to width with a table saw (for the shorter pieces) or a circular saw and rip guide (for the longer pieces). We used a thickness planer to dimension the handrail, ending up with a 2¼-in. wide by 2½-in. high cross section. To feed the curved stock through the planer with the concave side facing down, we built a jig to support the wood within the plane. First we screwed a length of half-round to a piece of plywood and then clamped the plywood to the planer bed so that the half-round was directly beneath and parallel to the cutter head. The half-round could then support the curved stock as we fed the stock through.

To match the existing molding profiles at the mansion, we ran a cove along the edge of the stringers and skirtboards and the rounded detail along the face of the trim. We routed the trim with our pin router, which I equipped with an automatic stock feeder. The skirts and stringers were too wide for the pin router, so we used a portable router instead, with a wide plywood fence screwed to the base for added support. The router bits we needed weren't stock items, so we ground our own from "blanks," which are common high-speed steel bits (for more on grinding router bits, see *FHB* #18, p. 75). We buy ours from Onsrud Cutter (P. O. Box 550, Libertyville, Il. 60048). They stock them up to 1½ in. high. Bits bigger than that can be special-ordered from some companies, such as Forest City Tool (P. O. Box 788, Hickory, N. C. 28603).

Though our stringers were curved, we approached their layout like we would for straight stringers, stepping off the spaces with dividers and scribing the cut lines using a framing square with stair-gauge fixtures attached (see pp. 8-14). We made the cuts with a worm-drive saw.

The handrail was shaped with the pin router, and we ground the router bits ourselves. It took four passes to complete the cuts. Though the finished depth of the railing was 2⅜ in., we planed it down to 2½ in. initially so there would always be a minimum of ⅛ in. of stock on the bottom to ride against the bearings on the router bits. Once we finished shaping the railing, we planed off the extra ⅛ in.

We also needed easements, goosenecks, and a volute to complete the railing. We glue-laminated stock for the easements and goosenecks, roughed them out with a bandsaw and shaped them with the pin router. Surprisingly, the pieces routed fairly easily, with just a little handwork necessary at the end. The volute was cut from a solid chunk of walnut.

Treads and risers—Meantime, we glued up quartersawn white oak lumber into 24-in. by 60-in. by 1⅛-in. thick pieces. Each blank was then marked with a plywood template, ripped on the diagonal into two treads and trued up

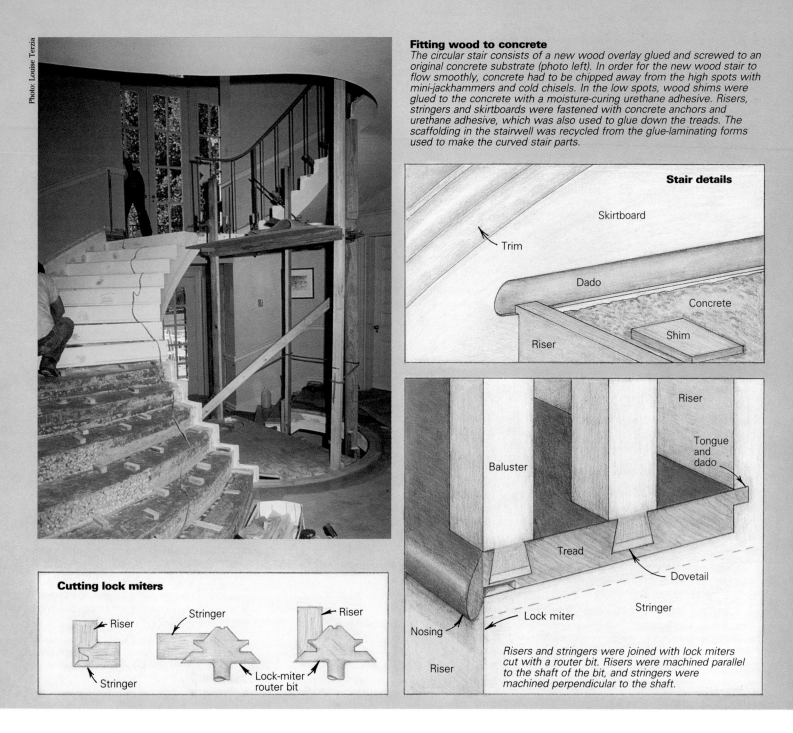

Photo: Louise Terzia

Fitting wood to concrete
The circular stair consists of a new wood overlay glued and screwed to an original concrete substrate (photo left). In order for the new wood stair to flow smoothly, concrete had to be chipped away from the high spots with mini-jackhammers and cold chisels. In the low spots, wood shims were glued to the concrete with a moisture-curing urethane adhesive. Risers, stringers and skirtboards were fastened with concrete anchors and urethane adhesive, which was also used to glue down the treads. The scaffolding in the stairwell was recycled from the glue-laminating forms used to make the curved stair parts.

Stair details

Skirtboard

Trim

Dado

Concrete

Riser

Shim

Riser

Baluster

Tongue and dado

Tread

Dovetail

Nosing

Lock miter

Stringer

Riser

Risers and stringers were joined with lock miters cut with a router bit. Risers were machined parallel to the shaft of the bit, and stringers were machined perpendicular to the shaft.

Cutting lock miters

Riser

Stringer

Stringer

Lock-miter router bit

Riser

---

on our jointer. Next, we measured in from the front edge of the treads the required nosing width (1¼ in.) and routed a ¾-in. wide by ¼-in. deep dado to house the tops of the risers (bottom drawings, above). On the rear of each tread, we routed a ¾-in. wide rabbet, which left a ⅜-in. square tongue to engage the riser above. After routing the bullnose on the fronts of the treads, we cut and mitered the inside ends to receive the return nosings, fitted a nosing to each and screwed them to the treads with 2¼-in. square-drive trim screws. Trim screws are like drywall screws with small heads, and are usually used to fasten wood trim to metal studs. The screws, which are driven with a small square-headed driver, allowed us to install and remove the nosings as many times as necessary to get perfect fits.

The risers were the easiest part of the stairway to make. All but the first six in the main

flight were straight, and they received only the ⅜-in. by ⅜-in. dado towards the bottom to receive the tread below. They were cut ¼ in. high so they'd slip into dadoes routed into the underside of the treads.

Cutting lock-miter joints—Because the inside stringer was open, a miter joint was required where the risers meet the stringers. I've always considered this less than ideal. It requires lots of glue and nails to hold it together, especially if there is no glue block on the inside corner as in this case. With newly available router bits, it's now feasible to cut lock-miter joints instead.

The beauty of a lock-miter bit is that it will cut both sides of the joint. It works by running one piece of stock parallel to the shaft of the bit and the other piece perpendicular to it (drawing above left). Pilot bearings don't work with this kind of router bit, so a fence is

required for either the router or the stock to ride against. For the stringers, we routed the miters from the backside, guiding each cut with a simple 1x fence clamped to the stringer a given distance away from the edge. We used a router table for the risers, running the stock vertically against the fence.

We drew the curved treads full size on plywood in our shop. The drawing allowed us to align the tread stock over it and to strike the various arcs with trammel points. The radius grew longer from the first to the last step, becoming a three-man operation just to draw (the largest radius was 200 in.). Once marked, the curves were cut on the bandsaw and then sanded on the edge sander and by hand. To cut the dadoes and rabbets, we used a router and a ¾-in. rabbetting bit fitted with a pilot bearing to follow the curves.

The risers for the curved steps were lock-mitered and dadoed like the rest. The material was birch plywood, kerfed on the backside so that the risers would bend and fit into the dado on the underside of the tread.

The Architectural Woodwork Institute (AWI) standards for circular stairs call for dovetailing the balusters into the treads. We were unable to find a manufacturer of dovetailed balusters, so we purchased the standard ¾-in. round-tenoned type and glued on a square block with a matching ¾-in. hole. We cut these to length and dovetailed them with a router and a shop-made jig. We used the same bit to route two dovetail housings in each tread.

Before moving our operation in our 16-ft. delivery van to the job site, all the stringers, skirts, trim and risers were painted with two coats of primer. The balusters and newels received two coats of finish paint, and we also applied two coats of transparent finish to the underside of the treads to prevent them from cupping later.

Concrete carving—Because the basement flight of stairs called for a lot of concrete removal, it was physically the toughest part of the whole job. We began by removing the carpeting, metal handrail and balusters (leaving the baluster anchors temporarily) and marking the high spots on the concrete steps with a felt-tipped marker. Then we attacked the high spots with a Bosch model 11306 demolition hammer fitted with either a 2-in. "wide chisel" bit or a bull point. The sound was incredible, reverberating up the well and all through the heavy brick walls of the mansion. Almost everyone, including the governor, found another place to do business for a couple of days.

The main flight in the foyer was also a bit off (photo facing page). Here, we decided to compromise by carving concrete away from the face of two risers with a cold chisel and planing off stock from the backs of the new risers. Any remaining discrepancies in both flights were dealt with during layout by slightly rotating the entire flight counterclockwise.

Concrete adhesive and anchors—Early on in the project, I contacted the Forest Products Laboratory (U. S. Department of Agriculture, Forest Service, 1 Gifford Pinchot Dr., Madison, Wis. 53705-2398) for some information about gluing wood to concrete. They made some suggestions, and we decided to conduct a test. We glued equal-size blocks of oak and poplar to the concrete floor of our shop, using several different adhesives, let them sit for several days and then tore them loose. The clear-cut winner was 3M's Scotchgrip 5230 wood adhesive, a moisture-curing urethane, which literally tore the wood off the sample. It also works with wood that has finish applied to it.

Because the stringers, skirts, trim and risers were all paint-grade, we could use plenty of concrete fasteners, too. We would simply putty the holes and paint over them later. We chose 3-in. sleeve anchors for the stringers

and 2-in. and 3-in. nylon nail-in anchors for the trim and risers (for more on concrete fasteners, see *FHB* #41, pp. 52-57). The main advantage of these anchors is that we needed to drill just a single hole through both the wood and the concrete. A hammer drill and carbide-tipped regular twist-fluted drill bits made short work of the drilling.

Installation—All the stringers were positioned and clamped in place until they could be secured. The skirtboards were fit initially by tracing the patterns onto them directly off the old skirts that we tore out. We then dropped them into position and wedged them against the wall using kickers off either the old steel balusters or off their anchors in the concrete. We made several trial fits, removing stock in small increments until we had a good fit. The stringer for the bulged wall in the basement fit fairly well, though we had to kerf the back to accommodate the bulge.

Before installing the skirtboards, we laid out the dadoes to receive the treads and risers (drawing top right, facing page). All the risers were 6-in. high, but the tread widths varied due to fluctuations in the outside radius. To determine these tread widths, we made a plywood template the same shape as a tread, but a couple of inches shorter. We screwed a block to the narrow end to hook over the inboard edge of the concrete treads and cut a hole so the template would clear the steel balusters. Then we dropped the template into position. By laying a straightedge along both the front and the back edges of the template and sliding the straightedge out until it butted against the skirtboard, it was possible to mark the width of each tread on the skirtboard. Once the layout was complete, we removed the skirtboard and routed the tread and riser housings with a ½-in. straight bit.

Where two stringers, skirtboards, or trim

Located in the foyer, the main newel post occupies a prominent position in the mansion. The top was enlarged and capped with a removable carving of the state seal. True to tradition, a dated scrap of paper hidden beneath it lists the names of all the craftsmen who worked on the project.

pieces joined at the junctures between stairs and landings, both were mitered at the same angle so their molding profiles would match up. To figure this angle, we adjusted a bevel square to match the angle between the stairs and the landing and drew the angle on a piece of wood. Then we bisected the angle with a pencil compass, adjusted the bevel square to the new angle and used it to scribe that angle onto the two adjoining pieces. Next, we cut the miters with our worm-drive saw and with handsaws, planing the miters with a block plane for a perfect fit. Once the cuts were completed, the pieces were glued on the back with 3M 5230 and wedged into position. Finally, they were drilled and anchored firmly to the concrete. The 5230 takes 24 hours to cure, so we didn't have to worry about the glue setting up before the stock was positioned properly.

Next, we measured the risers and cut them to length. After gluing shims to the concrete with 5230, we applied yellow glue to both ends of each riser, tapped the outboard end into its housing and the inboard end to engage the lock miter on the stringer. The miters were screwed or nailed tight, and the risers were anchored to the concrete with the 3-in. nylon anchors.

To fit the treads, we first cut the old balusters and anchors out of the way. Then we measured the finished length of the treads and cut the treads ½ in. longer than that to allow for scribing. We set them in place, scribed them to the skirtboards and then cut them with a bandsaw. The treads were glued to the risers with yellow glue and to the concrete (or to shims glued to the concrete) with 5230. For the curved steps, we screwed and glued the upper riser to the back of each tread first (for the bottom step, we also glued on the lower riser), then glued and anchored them to the concrete. An 8-lb. sledge hammer and a block helped us to persuade treads and risers into place.

For the landings, we bandsawed 1⅛-in. thick quartersawn white oak nosings, bullnosed them and glued them to the edge of the existing marble flooring with 5230. Then we mitered the skirt and trim pieces and glued and anchored them in place.

The treads were finish-sanded with 220-grit sandpaper, given two coats of stain and filled with a paste wood filler. Then they were finished with two coats of polyurethane. Once the balusters, newels and handrail were installed, we finished the handrail with several coats of natural Watco oil.

Winding up—The main newel post in the foyer intrigued me from the very beginning of the project. Located in perhaps the most prominent position in the whole mansion, it seemed to demand some special attention. And with the help of carver Paul Emrie, that's what we gave it (photo left). □

John Alexander is an architectural woodworker with Walnut Fork, Inc., Near Deer, Ark.

Building a Helical Stair

Laying out a spiral stringer with a little help from the trig tables

by Rick Barlow

A house I recently finished near Telluride, Colo., needed a spiral stairway to wrap halfway around its 4-ft. dia. stone chimney. I decided that a helical stringer supporting the treads by fabricated steel brackets would satisfy structural requirements and complement the house's contemporary design.

In theory, making a laminated wood helix isn't that difficult. Basically all you do is wrap successive layers of wood around a cylinder so that it spirals upward at a constant angle. But to glue up a helical bent lamination like the one I wanted for my stair stringer would require a cylindrical bending form larger than the chimney itself. I could have framed up such a form from curved plywood plates and 2x studding, but that would have eaten up a lot of time and materials.

I chose instead to make a bending form from five pieces of ¾-in. plywood and a strip laminated from three layers of ¼-in. plywood (drawing, below). The trick was to space the on-edge pieces of ¾-in. plywood at equal intervals along the proposed rise of the stairway, and to locate on each of these supports—or bulkheads—a 10-in. wide plane that would lie precisely ¾ in. below the surface of my theoretical half cylinder. To these five planes I could attach three layers of ¼-in. plywood. Glued together, this rigid plywood curve would be the platen, or base, of the bending form. To it I'd clamp the mahogany plies that would make the helical stringer.

In section, I wanted the stringer to be 5 in. wide by 6 in. deep, with its corners radiused with a ¾-in. roundover bit. Plies ¼ in. thick would easily make the required bend, and I used 20 of them 6 in. wide.

The centerline of the helical stringer would lie 42½ in. out from the centerline of the chimney cylinder. This meant that a typical 36-in. wide tread would clear the stonework by about ½ in. Subtracting half of the stringer thickness (2½ in.), gave me a 40-in. radius for the concave side of the stringer. This dimension then was the radius of the outside (convex side) of the bending form.

To create the helix, the first and fifth supports of the bending form needed to support the stringer against vertical edges. The second and fourth needed 45° edges, and the stringer would run over a horizontal edge at the middle support.

I nailed the five plywood bulkheads to the floor with 2x4 blocks and braced them diagonally to handle the forces they'd have to withstand during gluing and clamping. Next, I bent the three layers of ¼-in. plywood over these supports. The first layer was screwed and glued to the bulkheads, and the next two glued on top. I made sure that during this process the whole assembly stayed properly aligned throughout its length.

I built the form so that its centerline would be 5 in. off the floor. This way the stringer could run long at each end and be trimmed to fit the floor and rim joist later. I covered the plywood helix with a strip of visqueen so the mahogany wouldn't get glued to the form.

Laying out the stringer—The laminations for the stringer started out as 16-ft. long mahogany boards. I had them milled to a thickness of ¼ in. and ripped into 6-in. widths. To determine the length of these pieces, a little trigonometry was required.

First, I drew the stringer in elevation as if it were a straight piece of wood (drawing, facing page). The height of the triangle thus formed was the rise of the stringer. This is the total rise (distance from the first floor to the second) minus one riser height, as in a standard staircase. The total rise (110 in.) breaks down into 14 rises of 7.86 in. (about 7⅞ in.). Therefore the total rise minus one rise (7⅞ in.) for the top step (landing) results in a stringer rise of 102⅛ in.

The base of the triangle represents the distance traveled by the stringer in its run around the chimney. In this stair, the stringer winds through a 180° arc around a 42½-in. radius at the center of the lamination. Its run, therefore, is $\pi \times 42½$ in., or 133½ in.

The length of the stringer itself is the hypotenuse of this right triangle. The Pythagorean theorem $(a^2 + b^2 = c^2)$ tells us it is 168 in. long, just over 14 ft.

Since the outer plies travel larger radii than those at the center, they required slightly longer pieces of wood. I ordered the 16-footers to be sure of sufficient length.

The drawing of the flattened helix also helped me determine various angles. By finding the cotangent of the triangle ($\cot A$ = adjacent/opposite = 133.5/102.125 = 1.3072), I got the pitch angle of the stringer, 37° 25′.

The complement of this angle is 52° 35′. This is the angle that the stringer makes when intersected by a plumb line, an angle I needed

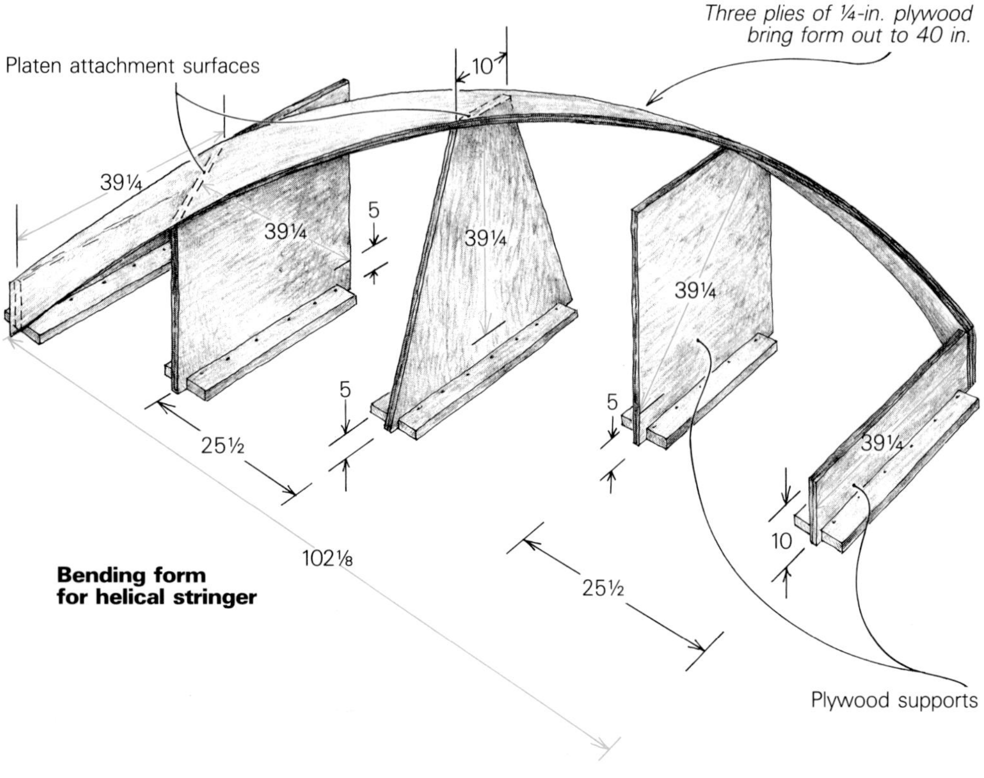

Three plies of ¼-in. plywood bring form out to 40 in.
Platen attachment surfaces
10
39¼
39¼
5
39¼
39¼
5
5
25½
39¼
10
102⅛
25½
Plywood supports

Bending form for helical stringer

Photos: Rick Barlow; Illustrations: Christopher Clapp

to know later when I counterbored the underside of the handrail to receive its posts.

I laid out the treads on the drawing of the straightened helix, and divided the 102⅛-in. rise of the stringer into 13 equal rises of about 7⅞ in. each. I figured the distance along the stringer from one tread to the next by dividing 168 (the stringer's length) by 13. This distance is 12.96 in.

Gluing up—Now, with the mathematics out of the way, I started gluing up the stringer. Since I wanted the glue to stick to the steel tread-support plates as well as to the wood, I used a resorcinol-formaldehyde glue. It's a hassle to use, and almost impossible to remove from tools and hands when it sets up, but it's strong stuff.

After spreading the glue with a small roller, I laid three or four 16-ft. long mahogany plies on the upright center support, clamped them there and then bent each side down to the ends of the plywood helix. You need lots of clamps. I have eight bar clamps, four handscrews, four smaller bar clamps and ten C-clamps. I borrowed about this many again from a friend and still had what I think is the minimum number of clamps for the length of this stringer, one clamp about every 3½ in.

Resorcinol glue has to cure at 70°F or above. It will set faster at temperatures higher than this, but it won't bond properly at temperatures below 70°F. Since I was doing this laminating in a house under construction in the wintertime at 9,000 ft. in the Colorado Rockies, I needed to camp out at the job site for seven days and nights to make sure the correct temperature was maintained while the laminating was going on. I glued up three or four layers at a time, letting them dry overnight before removing the clamps.

To rout the slots that would receive the steel tread-support plates, I laminated the stringer without glue between the third and fourth laminations and between the 17th and 18th laminations. This let me separate the first three and the last three plies from the center core of the stringer, rout out slots for the plates, insert and secure the plates, and then glue these inner and outer laminations onto the center one. The entire stringer was shaped and sanded before the plates were inserted so the steel would not get in the way.

I made two guide templates for my router to rout out the slots. One was for the thirteen slots on the convex curve of the stringer, and the other was for the thirteen slots on the concave curve of the stringer. Each jig was made to rout a 3-in. wide slot ¼ in. deep. To locate and orient each slot, I hoisted the center core lamination of the stringer into the position it would ultimately take, trimmed off the ends to fit, and temporarily secured it there

Barlow's helical stairway winds its way around a 4-ft. diameter stone chimney. The single stringer was laminated from 20 plies of ¼-in. mahogany. The treads, made of plywood and solid cherry end boards, are supported by brackets fabricated from ¼-in. steel plate.

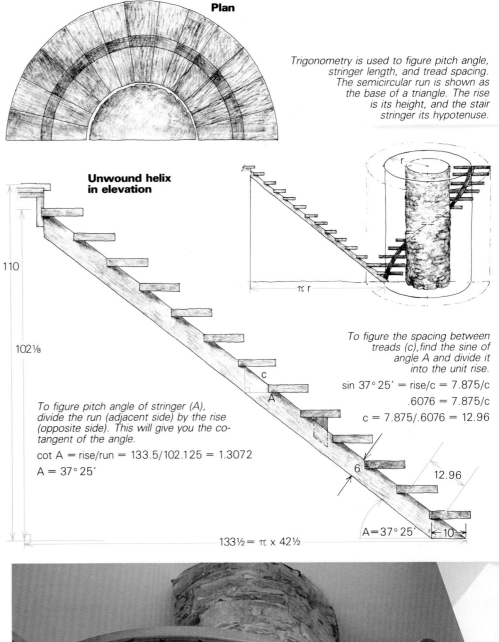

Plan

Trigonometry is used to figure pitch angle, stringer length, and tread spacing. The semicircular run is shown as the base of a triangle. The rise is its height, and the stair stringer its hypotenuse.

Unwound helix in elevation

110

102⅛

π r

To figure the spacing between treads (c), find the sine of angle A and divide it into the unit rise.

$$\sin 37°25' = rise/c = 7.875/c$$
$$.6076 = 7.875/c$$
$$c = 7.875/.6076 = 12.96$$

To figure pitch angle of stringer (A), divide the run (adjacent side) by the rise (opposite side). This will give you the cotangent of the angle.

$$\cot A = rise/run = 133.5/102.125 = 1.3072$$
$$A = 37°25'$$

c

A

6

12.96

A = 37°25'   ← 10 →

133½ = π x 42½

laminating two ½-in. thick pieces onto a ¾-in. thick core, yielding a total thickness of 1¾ in.—the same as the cherry. The ¾-in. middle ply of plywood was cut 1 in. shorter on each end, to form a ¾-in. by 1-in. groove. I cut tongues on the cherry end boards and screwed and glued them to the center pieces.

I bandsawed the ends of the treads to conform to the semicircular plan view, and rounded over all the edges with a ½-in. pi-loted router bit. Finally, I fastened each tread to its mounting plate with four lag bolts.

Bending the railing—The form I built to laminate the handrail was like the one I used for the stringer, only with a larger radius. The handrail would be just under 16 ft. long, so I could use the same boards. I used eight laminations ¼ in. thick and 3 in. wide, making the handrail 2 in. by 3 in. in section. To accept the handrail posts, I counterbored two ½-in. dia. holes in each tread at the centerline radius of the handrail, and spaced them evenly 7¾ in. apart. I then counterbored the underside of the handrail. These holes had to be drilled at an angle of 52° 35′ (the complement of the pitch angle). To find the proper spacing for

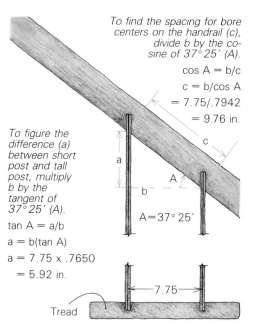

To find the spacing for bore centers on the handrail (c), divide b by the cosine of 37° 25′ (A).

$$\cos A = b/c$$
$$c = b/\cos A$$
$$= 7.75/.7942$$
$$= 9.76 \text{ in.}$$

To figure the difference (a) between short post and tall post, multiply b by the tangent of 37° 25′ (A).

$$\tan A = a/b$$
$$a = b(\tan A)$$
$$a = 7.75 \times .7650$$
$$= 5.92 \text{ in.}$$

A = 37° 25′

Tread

the bore centers here, I divided 7¾ (the horizontal spacing) by the cosine of the pitch angle, as explained in the drawing. Then ½-in. dia. steel rods, painted flat black, were inserted in the treads. Since each tread gets two rods, one must be longer than the other. By multiplying the tangent of the pitch angle by 7¾, I found the regular difference in length between the long and short rods.

With the help of friends, I worried the rods into their holes one at a time, starting at the bottom rod and working up to the top. By using a rubber mallet and some clamps, I brought the handrail down onto the 26 rods.

All that remained was to carpet the plywood center pieces and metal tread plates. Since the treads are visible from below, this was a difficult job, but it turned out fine. □

Rick Barlow is a contractor in Colorado.

Before its treads are attached to the steel-plate brackets, the stairway is installed. The temporary midway support will be replaced by a steel pipe that will be grouted into a masonry planter below.

with clamps and a 2x4 leg at its midpoint. Then using a level, I drew plumb lines on each side of the 26 slots. This way I was sure they would be plumb in the finished staircase.

These slots were 3 in. wide, and their sides followed the curve of the stringer, so the legs of the plates also had to be slightly curved. I hit the middle of each leg with a sledge while it rested on two other plates along their edges. One or two whacks did the trick.

I drilled holes in these plates for four 8d nails to hold them onto the stringer. These and the resorcinol-formaldehyde glue made for a solid bond between metal and wood.

With the tread-support brackets installed, I put the stringer back onto the form and glued the inner and outer laminations onto the core. They were already shaped and sanded, so I had to align the pieces precisely, and use pine blocks to keep the soft mahogany from being crushed by the clamp jaws. I was also careful to spread the glue so not much would ooze out of the joints. What squeeze-out there was, I wiped off before it dried. This assembly was left to dry overnight (photo above).

Installing the stringer—After a little more finish sanding, I cut the outer plies to fit the floor and rim joist. A 180° helix like this one

needs support at its midpoint. I drilled a 2-in. hole underneath the stringer and inserted a galvanized pipe that sits discreetly in some rockwork below.

When the stringer assembly was firmly attached in place, the horizontal tread plates were welded to the vertical plates protruding from the stringer. Using a universal level, I held the horizontal plates level in both planes, while a welder attached the plates from underneath. After cleaning up the welds with a file and wire brush and before attaching the treads, I oiled the stringer and painted the plates and the support pipe flat black.

Making the treads—Working out the dimensions of the tapered treads required more figuring. The radius to the outside of the treads was 60½ in. Multiplying by π gives the length of a 180° arc, or 190 in. Dividing this by 13 treads gives 14.62, so the outside arc of each tread is 14⅝ in. Similarly, I found the length of the inside arc of each tread to be 5¹⁵⁄₁₆ in. Nosing added an inch, so each tread is 36 in. long with a 15⅝-in. arc at its outer end and a 6¹⁵⁄₁₆-in. arc at its narrow end.

I planned to carpet the centers of the treads, and leave solid cherry end boards on each end. I made a center of AC plywood by

Double-Helix Stair

A cherry spiral built with divine inspiration, a little steam and a lot of clamps

by Robin Ferguson

The house is majestic, sitting on a bend of Castle Creek, 8,000 ft. up in the heart of the Colorado Rockies. And a castle it seems, with its medieval lines and three towers. The central, and highest, tower was designed for solitude, meditation and glimpsing the sunset. When the clients looked at the architect's model of the house, they liked the tower. It was just what they wanted, but with the tower centered over the living room, they wondered how they'd get up there. In response, Steven Conger and Michael Martins, who had designed the house, conceived a sculptural, double-helix stair (photo right).

The design and construction of the stair was a group effort that evolved over many months. On hearing the architects' concept for the stair, construction foreman Ivar Eidsmo said he'd seen one just like what they had in mind. He was referring to the Miraculous Staircase in the Loretto Chapel in Santa Fe, N. Mex. (photo below). It was built about 1878 by an itinerant carpenter who appeared out of nowhere, apparently in re-

Unlike conventional spiral stairs that are supported by a central post, the cherry staircase at right stands on two helical stringers. It rises over 16 ft. to an observation tower above the living room. The design of this double-helix stair was inspired by the Miraculous Staircase (above), which leads to the choir loft of the Loretto Chapel in Santa Fe, N. Mex.

Alan Becker

Scarf-joint detail

Scarf joints were cut on an industrial table saw at a local millwork shop, and staggered in stringer (right) to maintain strength.

Plan view of column and cage

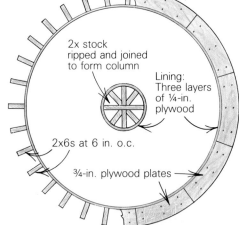

2x stock ripped and joined to form column

Lining: Three layers of ¼-in. plywood

2x6s at 6 in. o.c.

¾-in. plywood plates

sponse to prayers to St. Joseph, carpenter of Galilee. The anonymous carpenter worked four months and disappeared as mysteriously as he had arrived, without finishing the railing and without payment for his work.

Eidsmo went down to Santa Fe and photographed the Miraculous Staircase. Although the stair we built is very different structurally and is shorter, narrower and much less ambitious, the Miraculous Staircase did serve as inspiration.

Rising over 16 ft. around a 2½-ft. radius and revolving 1½ turns along the way, our stair was a major undertaking. It was begun by Chuck Miller, one of the carpenters working on the house. I came along later, after the layout had been done and the forms built. Initially my job was to speed up production, but when Miller had to leave the project, I took responsibility for its completion.

Specifications—During the design stage, an engineering firm, Nicol and Giltner, was hired to do structural specifications for the stair. To match the interior woodwork, the staircase is made of cherry. So the engineers based their calculations on the strength of that wood.

They determined that the stringers should be 2 in. thick, composed of eight layers: four ¼-in. plies in the middle, two ⁵⁄₁₆-in. crossband plies with their grain running perpendicular to the others, and two ³⁄₁₆-in. face veneers.

Because of the wedge-shaped treads, the inside stringer needed to be 7 in. wide, while the outer one had to be 14 in. wide. At the top and bottom of each stringer, a ¼-in. steel plate had to be sandwiched across the full width and securely lag-bolted to the landing and floor system. To hide these plates and cover the various structural laminations, we had to glue 1-in. thick caps, made with eight layers of solid cherry, to the upper and lower edges of the stringers.

In addition, the engineering firm also specified the 2-in. thickness of the treads and their installation details. The treads are held in place by five ½-in. dia. lag bolts 6 in. long—two through the inside stringer and three through the

outside one. The heads of the bolts are concealed beneath the final ply.

Since the stringers are so long (the outside one is nearly 28 ft.), each ply is composed of sections, 6 ft. to 8 ft. long, glued up to form one full-length piece. For maximum strength, the adjoining ends of each ply were scarf-jointed (drawing, top left) on a huge industrial table saw at a local millwork shop.

Column and cage—Building the laminating forms was the first big hurdle in the construction process. Rather than build the stair in a woodshop and then transport and install it, we set up the laminating forms right in the stairwell. The stringers were laminated in place, and the forms were then stripped away from them.

We built an inner column of solid 2x fir, and an outer cage encircling it (drawing, middle left). These we wrapped with plywood along the path of the stringers. The cage consisted of vertical 2x6s on 6-in. centers. We used ¾-in. plywood to make the curved plates by bandsawing segments to form a 5-ft. dia. circle.

The column was made with seven pieces of 2x stock stood on end and cut to radiate out from a central axis, like spokes on a wheel, to create a star-shaped cross section with a 12-in. diameter. Three layers of ¼-in. AC plywood were used for the lining, and this is where our wood-bending difficulties began.

Straight curves are relatively easy to bend, but compound or twisting curves put a much greater stress on the wood, so we had to use steam. Eventually we built three different steamboxes, each one longer than the last. These were simple plywood boxes (6 in. by 14 in.) with a perforated copper pipe running through them, and fed by a 2½-gal. kettle (photo below left). Steaming made the plies supple enough, but once out of the box, the thin wood cooled so fast that we had only seconds to get the piece bent in the form. With the help of several people, we held it there and screwed it to the form with plenty of drywall screws.

As each full-length layer was completed, careful truing had to be done before the next could be added. We checked the curves with templates cut to the appropriate radius (which changed with each layer). Any high spots caused by the steaming or by slight imperfections in the column or the 2x6 cage were removed with a lot of grinding and hand sanding. This was a tedious process, and given the length of the forms and the cramped 2-ft. space between column and cage where we were working, we came to appreciate the term "cage."

Clamping setup—The engineering firm we consulted had specified resorcinol as the best glue for our project. Commonly used in plywood and gluelam beams, resorcinol requires extreme clamping pressure (175 psi) to bond properly. To satisfy this requirement, we used vertical 2x6s on edge, forming a continuous row of cauls or battens along the length of the outside stringer. We ran ⅜-in. threaded rods through the ends of the cauls and into steel plates that had been tapped and lagged securely to the form (drawing, facing page, top right). Now we

The twisting curves in this stair are so severe that even the thin laminations had to be steamed before they were pliable enough to bend around the forms.

could apply plenty of pressure and distribute that pressure uniformly.

Before starting on the stringers, we laminated a pressure board from four layers of ¼-in. plywood to place over the stringer during the gluing operation to protect it and further distribute the clamping force of the cauls.

Before we could begin laminating stringers, we built another steambox, this one 30 ft. long, to accommodate the veneers. Once we got started, we soon learned that oversteaming did no good. After about 20 minutes, the cherry veneers were as pliable as they were going to get. Extra time only made the wood swell—a potential disaster when it shrank later.

As with the plywood, the cherry plies cooled rapidly, and getting the pieces from the box to the form in time was hard. Four people would run the plies into the house and feed them up to four others positioned around the form. Cherry isn't the best wood for bending, but there could be no dry runs—glue was rolled on to the previous lamination, the ply run in and clamped.

Resorcinol sets in about two hours—so the directions say—but here in the Rockies, the humidity is very low. Getting the ply in place, pressure boards on and all 90-odd clamps tightened before the glue set was frantic work.

On the inside stringer, we couldn't use the same clamping technique because the radius was too tight and the rise too steep. Instead of full-length plies, we laid up sections and clamped them with metal straps and a Signode model DO-3A banding machine (Signode Corp., 3610 W. Lake Ave., Glenview, Ill. 60025), like those used in lumberyards to bind stacks of plywood. Using about 30 bands around the column for a 5-ft. length, we achieved tremendous pressure. This method worked very well. As with the form itself, each full-length layer was carefully trued before the next was applied.

Adding the caps—We could laminate only six of the eight layers around the form. So after laminating as many as we could, our next step was to square the upper and lower edges of the stringers and glue on the caps. For this we used a 1½-in. straight-flute bit in a router. To keep the bit from wobbling on the curved surface, we had to sculpt wooden bases for the router, both convex and concave, to match the stringer's shape.

We attached a flexible guide strip along the face of the stringer for the router to ride against. As with most of the work on this stair, there was no convenient way to arrange a trial run. So the routing itself was a very intense process; we were keenly aware that an errant cut couldn't be easily hidden or fixed.

To apply the cap strips, we had to rethink our clamping methods, for pressure was now required in two directions: against the form and onto the stringer. We drilled 3-in. thick cherry blocks and threaded bolts through them. With the blocks screwed to the form, the bolts could be run down to exert pressure on the cap (drawing, bottom right). Even though we were using machine-thread bolts, the wooden threads they cut in the hardwood blocks were plenty strong enough for our purpose.

For holding the cap strips against the form,

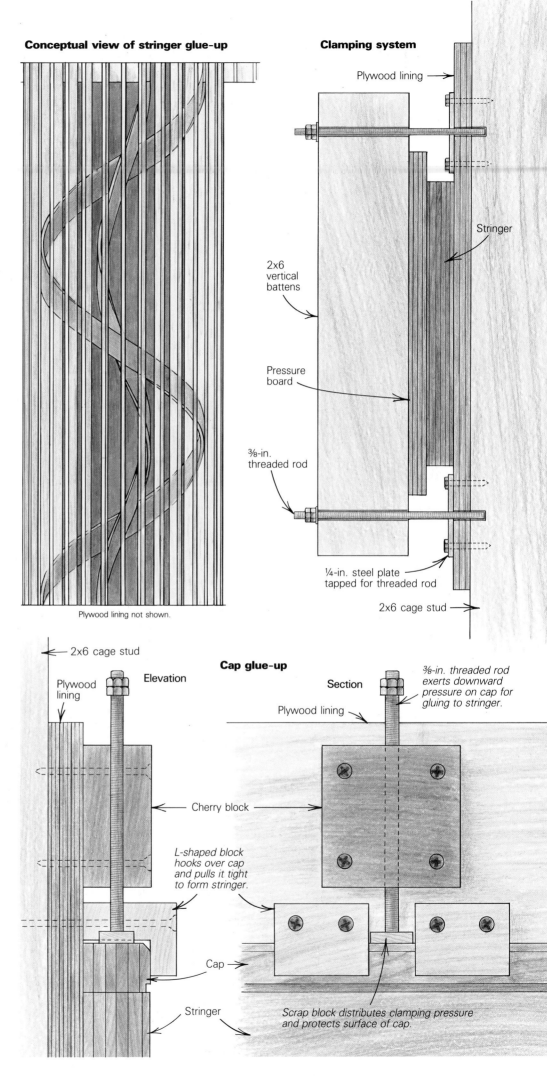

Conceptual view of stringer glue-up

Plywood lining not shown.

Clamping system

Plywood lining

2x6 vertical battens

Pressure board

⅜-in. threaded rod

Stringer

¼-in. steel plate tapped for threaded rod

2x6 cage stud

Cap glue-up

2x6 cage stud

Plywood lining

Elevation

Cherry block

L-shaped block hooks over cap and pulls it tight to form stringer.

Cap

Stringer

Section

⅜-in. threaded rod exerts downward pressure on cap for gluing to stringer.

Plywood lining

Scrap block distributes clamping pressure and protects surface of cap.

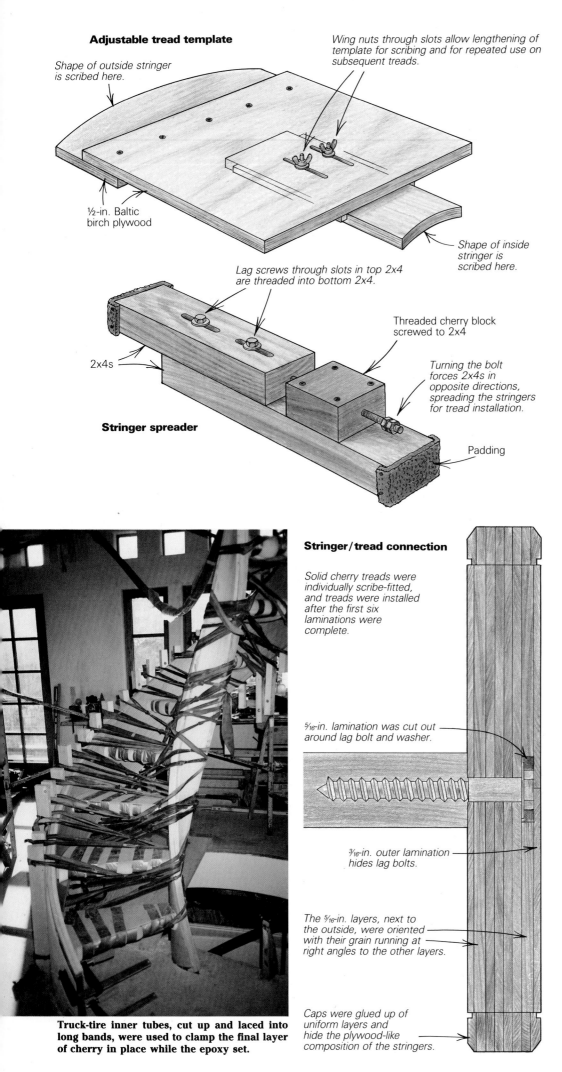

Adjustable tread template

Shape of outside stringer is scribed here.

Wing nuts through slots allow lengthening of template for scribing and for repeated use on subsequent treads.

½-in. Baltic birch plywood

Shape of inside stringer is scribed here.

Stringer spreader

Lag screws through slots in top 2x4 are threaded into bottom 2x4.

2x4s

Threaded cherry block screwed to 2x4

Turning the bolt forces 2x4s in opposite directions, spreading the stringers for tread installation.

Padding

Truck-tire inner tubes, cut up and laced into long bands, were used to clamp the final layer of cherry in place while the epoxy set.

Stringer/tread connection

Solid cherry treads were individually scribe-fitted, and treads were installed after the first six laminations were complete.

⁵⁄₁₆-in. lamination was cut out around lag bolt and washer.

³⁄₁₆-in. outer lamination hides lag bolts.

The ⁵⁄₁₆-in. layers, next to the outside, were oriented with their grain running at right angles to the other layers.

Caps were glued up of uniform layers and hide the plywood-like composition of the stringers.

we cut 2-in. wide L-shaped blocks that hooked over the edge of the cap. We spaced them every 4 in. and pulled them tight to the form with drywall screws. Once the cap strips were secure, we no longer needed the forms and set about to dismantle them. All the drywall screws did their best to keep the column and cage intact, and it seemed for a time that the stringers would be destroyed with the forms.

Treads—The main purpose of all the careful truing was so that one tread pattern could be used for the whole stair. But so much for careful planning. With the forms stripped away, the stringers took a set of their own, and while the differences were minute, a single tread pattern was now out of the question. We were obliged to cut 21 individual treads, each having to fit perfectly between the convex and concave surfaces of the stringers.

In order to make individual patterns for each of the treads, I built an adjustable template of ½-in. Baltic birch plywood (drawing, top left). I made it in two pieces, held together with bolts and wing nuts in oblong slots, and cut each end to the appropriate radius. It could then be extended to make contact with both stringers. After scribing the exact curve, I belt-sanded down to my line. I repeated the process (each time opening the jig slightly) until I had a perfect fit. The template was then used to make a pattern of ⅛-in. tempered hardboard. We glued each pattern to a tread blank, bandsawed to within ¹⁄₁₆ in., and trimmed it flush with a 2-in. bearing-over router bit.

The treads were such a tight fit that they couldn't be installed without marring the faces of the stringers. We had to make another device to spread the stringers slightly so we could get them into place (drawing, middle left). For this I used two pieces of 2x4, held together with lag bolts run through slots, and just tight enough to hold the two together yet allow movement. With one of the blocks of cherry from the cap glue-up (tapped for a bolt) mounted on one of the 2x4s, and the bolt against the end of the other, the two could be forced apart by turning the bolt.

Having installed all 21 treads, we were able to use the stair. It was quite a sensation, after so many months, finally being able to climb the thing. Even so, some of the most challenging work still lay ahead.

Epoxy to the rescue—Applying the last two plies required still other methods of clamping. Since we had dismantled the forms, our old system was no longer available to us. Satisfying the high pressure requirements for resorcinol would have dictated major retooling.

I knew about epoxies but had never used them for anything but plastic laminates. But for the final plies they proved a great problem solver. Instead of resorcinol, we used epoxy that had an open time of six hours, with a full cure in seven days. Rather than clamp each ⁵⁄₁₆-in. layer, we stapled it in place, since contact was all the epoxy needed for a strong bond. Before clamping, we cut holes in the wood around the lag bolts, and after the epoxy had set, ground the heads flush. These holes in turn were filled with

Photo: Michael Owsley

auto-body filler and sanded smooth. It was time at last for the face veneer of cherry.

We couldn't use staples on the last veneer so our clamping system changed again. To hold the veneer tight to the stringer during glue-up, we used a series of 2x2 vertical cauls, snugged against the veneer with 30 truck-tire inner tubes. After cutting them into rings and lacing them into bands, we wove the tubing between the treads and around the opposite stringer (photo facing page). This worked amazingly well.

Railing and connections—The handrails had been glued up before the forms were dismantled, but they needed their upper and lower surfaces squared before they could be installed. Trying to work the outside rail required a great deal of patience. At nearly 28 ft. long, 1½ turns and a 5-ft. diameter (photo top right), just getting it clamped was like wrestling a giant snake. Once it was clamped, we could work only a short portion before the twist had us working upside-down and we were forced to reposition it.

We hand-planed the surfaces of the rail, trying to follow the twist with each stroke. It was impossible to avoid some tearout because, with all the plies, there was always grain running in opposite directions. But we cleaned this up with sharp cabinet scrapers.

Once shaped, the rails were clamped temporarily in place. Because of slight variations in the distance between stringer and rail, the square balusters had to be fit individually. They are doweled into both the handrail and stringer.

One unusual feature of this stair is that it has two handrails. A regular spiral stair built with a center post has only an outside handrail. Our stair also has a rail that hovers just above the inside stringer, so you can hold on with both hands as you go up or down. Two rails make the stair more comfortable and secure to use.

Foam patterns—The final phase, and for me the most enjoyable, was designing and carving the many rail connections. At the suggestion of Michael Owsley, who aided greatly in the completion of this stair, we glued up pieces of polystyrene foam into big chunks. We wanted to experiment with these before making the transitional rail connections in cherry.

After permanently installing all the long sections of rail, we glued the polystyrene chunks in place between the rails with five-minute epoxy and a piece of paper sandwiched between the foam and wood. (The paper allowed us to break apart the glue joints later.) These were then shaped with various saws and rasps to form the final models (photo bottom right).

Passing inspection—The Uniform Building Code says that a staircase shall not rise more than 12 ft. vertically between landings. Our staircase rises over 16 ft., but we were granted an official variance by the building department. Since this is a low-use stair that leads only to an observation tower and is not for egress, the inspector felt the variance was justified. □

Robin Ferguson is an architectural woodworker in Snowmass, Colo.

Top photo: Germaine Dietch; Bottom photo: Robin Ferguson

Maneuvering the 28-ft. long handrail into place was like wrestling a giant snake (top). Once it was installed, the builders glued foam blocks in place temporarily (above) to work out the shapes for the transitional sections of handrail.

Building a Winding Outdoor Stair

Fitting stairs to a complicated site called for an accurate sketch and some job-site improvising

by Thor Matteson

A couple of summers ago, I was helping my father build a house in El Portal, California, when some neighbors, Ron and Liz Skelton, asked if I'd be interested in building a deck for them. They offered to *pay* me for my work; my father understood the lure of fortune and fame and encouraged me to accept the job, which I did.

The Skeltons wanted a redwood deck to replace a sagging fir assembly built over a rock garden. That part of the job would be tricky enough, but there was more. Ron is the local mechanic, and he had no stairs leading from the house down to his shop located one level below. After several years of skidding down the bank to work, he was ready for a stairway. Ron described to me where he and Liz envisioned the stairs—"starting about here, winding down between these rocks and ending up at the slab in front of the shop," he narrated while scampering down the bank around several boulders. Suddenly I felt slightly dizzy. I was barely getting the hang of *straight* stairs, let alone double-inflected variable-radius curved stairs. Nevertheless, I forged ahead with the deck and extended a landing from which to start the stairs.

Liz and Ron wanted a sturdy-looking stair, so we settled on 42-in. wide (approximately) open-string stairs with 2x12 stringers and 2x6 risers, both of redwood, and 4x redwood treads (photo facing page). Four-in. thick redwood can be tough to find, and we needed 4xs for the deck framing, too. Luckily, a local contractor gave Ron a tip on where to get it, and the supplier sent the most stunning redwood I'll probably ever see—vertical grain, surfaced full-dimension 4x6s and 4x8s, 16 to 20 ft. long with hardly a defect in them. I saved the nicest stuff for the treads and sentenced the rest to life beneath the decking.

Measuring and sketching—When building an unconventional stair such as this one, it's a good idea to start with a drawing (drawing facing page). I mapped out a plan view first, scaled ½ in. to the foot and identified the critical elevations. That called for a few hours scaling the hillside with a tape measure, a plumb bob and a builder's level.

To figure out the total rise of the stair, I perched the builder's level near the top-land-ing location and shot the elevations of both the top landing and the shop slab (the bottom slab of the stair would butt up level to the shop slab). Then I subtracted the top elevation from the bottom and came up with a total rise of 127 in. I don't have a leveling rod to use with my level, but a tape measure attached to a straight 2x4 served me just as well.

Next, using framing members of the new deck as reference points, I measured horizontally and dropped a plumb bob to locate the relative positions of the three boulders around which the stairs would wind, plus the location of the bottom landing. I used all of these measurements that evening to draft a plan view of the entire project. The path the stair would have to take was now obvious, and I penciled it into the drawing.

Now I drew in a line which I thought best represented the line of travel someone was likely to take in traversing the stairs, picturing myself swinging around the end post on the deck railing and trotting down the steps, brushing past the boulders while taking the shortest route to the shop. Then I laid out the stair so that the rise and run of each step would be consistent along that line of travel. I walked my architect's scale along the curved path line, and I decreed the scaled length of the path line to be the total run of the stairs—20 ft. even (by coincidence). I converted this figure to inches and used this number, along with the 127-in. rise to figure out the rise and run of each step. The rule of thumb for stairs states that rise plus run should be between 17 and 18 in. and that rise times run should equal about 75 in. I decided on a rise of 6³⁄₆₄ in. and a run of 12 in., which added up to 21 risers and 20 treads. Though the rise and run weren't quite what the time-honored formulas sanctioned, in the end they worked just fine.

Now I knew how many treads to put into my sketch, but I still had to figure out their orientation and shape. This I did mostly by instinct. I put a tick mark every 12 in. along the path line in my sketch and used the marks as points on which to pivot the tread nosing. I simply tried to keep the taper of consecutive treads from changing too abruptly.

Once I was happy with the tread layout, I sketched in stringers as close to the ends of the treads as possible to avoid excessive can-tilever. It took five pairs of stringers to skirt all the obstacles, and each of them was of a different length and slope (in any circular stair, inside stringers are steeper than outside stringers). The shortest stringers would carry three treads and the longest stringer would support seven treads.

Placing the piers—I knew where to place the piers to support the stringers, but I still had to determine their elevations. To do this, I calculated the elevations of all the treads above the shop slab and labeled them on my drawing. Then I drew a cross section of each tread and stringer that fell directly over a pier and figured as best I could the distance between the tops of the treads and the tops of the corresponding piers (the slopes of the stringers varied, so these measurements varied). By subtracting these numbers from the tread elevations, I obtained the heights of the piers.

With the drawing complete, it was time to return to the site and locate the piers, again using the deck framing as the point of reference. Then I shot the elevation of each one with the builder's level. I soon discovered that I was in for considerable digging. The required scraping simultaneously cleared the path for the lower two-thirds of the stairway and contributed fill to expand Ron's parking area.

Once I was convinced that the calculated pier elevations were correct, I poured the concrete piers and oriented wet post anchors in them to accept the stringers.

Solving the stringers—When the concrete set up, I began to work on the stringers. I worked from top to bottom so I could compensate for accumulated errors by adjusting the finished level of the concrete slab at the base of the stair (the accumulated error amounted to less than ⅛ in). The usual framing square and stair-gauge fixtures were useless for the layout of these stairs because of the different slopes of the stringers and because nearly every tread was a different shape. Instead, I laid out the stringers in place.

I started out by making plumb and level cuts on the ends of the upper pair of string-ers, figuring the cuts by taking the rise and

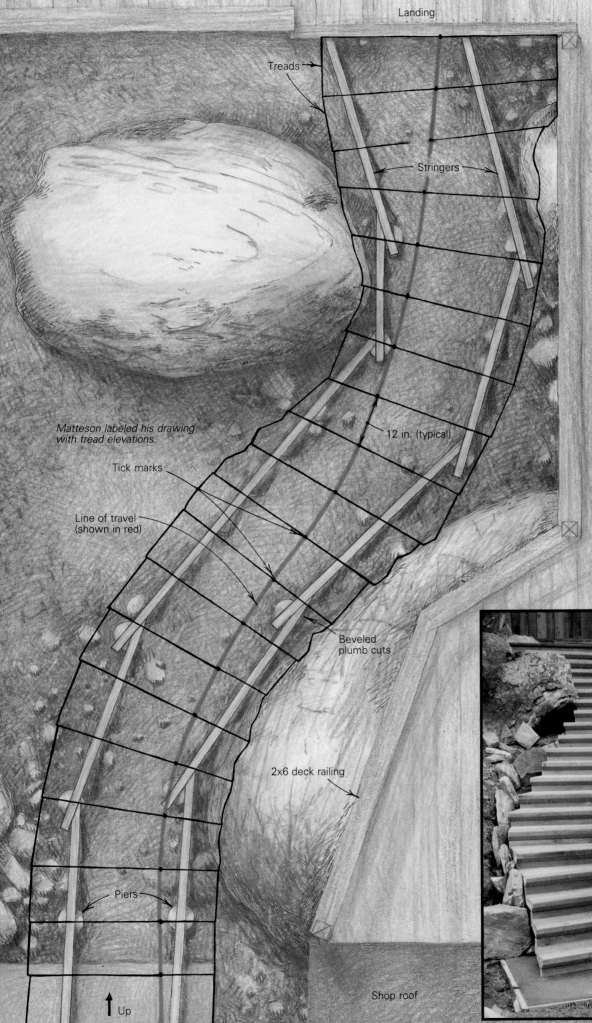

Landing

Treads

Stringers

Stair layout

The meandering stair connects the entry-level deck to the slab in front of the mechanic's shop below. Most of the stair layout was calculated on paper. For his drawing, Matteson first rendered the existing deck, landing, shop slab and boulders to scale, and sketched in the rough outline of the stair. Then he drew a line to represent the theoretical line of travel. He walked an architect's scale along the line and measured a total run of 20 ft., or 12 in. per tread. By making a tick mark every foot (to scale) along the line of travel, he could then use the marks as pivot points for the treads, drawing and erasing the nosings until the treads followed the curvature of the stair gracefully. Matteson then sketched in the five pairs of stringers required to skirt the obstacles, drew in the piers to support them and labeled his drawing with the tread elevations. Pier elevations were derived from the tread elevations. On-site layout was then scaled off the drawing, and the elevations of the piers were shot with a builder's level.

Matteson labeled his drawing with tread elevations.

12 in. (typical)

Tick marks

Line of travel (shown in red)

Deck

Beveled plumb cuts

2x6 deck railing

Piers

Up

Shop roof

Shop slab

run of each stringer off my drawing. After notching the bottoms slightly to clear the piers, I propped the stringers against a 4x6 header fixed to the landing and clamped them to the post anchors at the bottom. Next, I marked the inside of the stringers 10¾₄ in. down from the deck surface (the rise plus one tread thickness) and used my 2-ft. level to scribe a horizontal line through each mark. Going back to my drawing, I scaled off the run for the first tread at both stringer locations (the run differed on opposite ends of the tread) and marked their lengths on the horizontal lines. I used my rafter square to scribe a 6³⁵₆₄-in. long plumb line down from each mark, then drew another pair of horizontal lines, repeating the process until I reached the end of the stringer.

In order for the risers to lay snug against the stringers, I needed to bevel most of the plumb cuts. To figure these bevels, I simply extended my plumb lines up to the tops of the stringers, laid my 4-ft. level across the tops and aligned the edge of the level with a pair of plumb lines. Then I scribed a line across the tops of both stringers. That gave me the angles to which I adjusted my saw for each cut. I unclamped the stringers and cut along the lines with a worm-drive saw, finishing the cuts with a handsaw.

Next, I replaced the stringers, drilled pilot holes in them and screwed them to the deck posts and to the post anchors with #8 zinc-plated bugle-head drywall screws, greased with a dab of beeswax from a toilet gasket. I had special-ordered 2,000 of these screws for the deck from Mid-Valley Distributors (3886 E. Jensen Ave., Fresno, Calif. 93725). They worked so well that I decided to use them for the stairs, too.

I approached the rest of the stringers the same way as the first pair, except that the subsequent stringers each required a beveled plumb cut at the top end where they met the preceding stringers. To lay out the cuts for a pair of stringers, I notched the stringers to fit the piers and laid them in place, snugging the top ends up alongside the upper stringers. Then I used my 2-ft. level to draw plumb lines at the joint locations and used a saw protractor to figure out the proper bevels, transferring the bevels onto the stock. Once the stringers were marked, I took them down and put my worm-drive saw to work.

The required bevels for the top cuts were all greater than 45°, so I couldn't cut them in a single pass. Instead, I made the plumb cut with the saw set at 90°. Then I set the saw to 90° minus the bevel angle and made the finish cut by resting the base of the saw on the end grain exposed from the first cut. This was a little awkward, but it worked. Next time, maybe I'll clamp a block to the stock for added base support.

The varying width of the treads had two unavoidable consequences. Sometimes the treads dropped faster than the stringers, leaving less meat in the stringers than I liked. To stiffen up these skinny stringers, I thickened them with a pair of 2x6s fastened with con-

struction adhesive and screws. On some of the outside stringers, I encountered the opposite problem—the stringers descended faster than the treads, leaving a triangular void between the riser, tread and stringer (photo facing page, right). I addressed that problem by gluing and screwing vertical 2x6 blocks to the stringer to provide extra support for the front corner of the treads.

After cutting all the stringers to receive the treads, I screwed them to the post anchors and to each other, occasionally checking my progress down the slope with the builder's level.

Treads and risers—Starting again at the top of the stairs, I began to install the treads and risers. Each tread butts up against the riser above it and laps the riser below it.

Most of the treads were composed of three pieces: a 4x8 nosing piece, a 4x6 center piece (usually with an angle cut along the rear to fit against the riser) and a small wedge to fill out any remaining space along the back. I cut the nosing piece of the top tread to length first, ripped a 45° chamfer under its front edge so it wouldn't look too chunky, and smoothed out the chamfer with a power plane. Then I set the piece on the stringers so that the bottom of the chamfer would be flush with the front of the riser to be installed later below it, creating a 1¾-in. nosing.

With the nosing piece positioned on the stringer, the next step was to fit the middle piece (drawing facing page). To figure out the taper on the middle piece, I first held my tape measure perpendicular to the rear edge of the nosing piece and marked the point along that edge where it measured exactly 6 in. to the riser. That point would locate the beginning of the taper on the middle piece. Next, I measured the perpendicular distance between the nosing piece and the riser at its narrowest point. I transferred these measurements to the blank for the middle piece and cut the taper with the worm-drive saw. This method gave me a cutting line more easily, quickly and accurately than measuring the angles could have, and I repeated it for the wedge piece.

When the three pieces of the tread fit tightly in place, I reached underneath and scribed a pencil line along the outside edges of the stringers. Then I removed the pieces and eased their top edges with a block plane. I turned the pieces upside down and screwed a pair of 1½-in. by 1½-in. cleats to the two front tread pieces. The cleats were positioned along the pencil lines, and would later be screwed to the stringers horizontally to anchor the treads. I used redwood shims between the cleats and the tread pieces to account for the varying stock thickness.

Next I stood the partly assembled tread on its nose and further connected the two front pieces with three ½-in. by 6-in. lag bolts, counterbored into the back edge of the middle piece deep enough so that the entire threaded portion of the bolt would grip the

nosing piece. For all this hole-shooting, I rigged up a few drills: a Milwaukee Hole Hawg with an Irwin expansion bit (set at 1½ in.) for the counter boring and a ½-in. Milwaukee "Magnum" drill for the lag-bolt shanks and pilot holes, plus I had my dad's antique clunker for drilling all the pilot holes for the drywall screws. The Hole Hawg is frighteningly powerful; a few weeks earlier it snapped the ⁹⁄₁₆-in. shank of a self-feed auger in mid-hole with hardly a twitch. The Magnum drill is nice and sturdy. The only problem is that the handle is mostly switch, and it's hard to grip without turning it on. Once the holes were drilled for the lag bolts, I drove the bolts halfway home with an impact wrench. The last few turns went faster with a ratchet wrench, which also allowed me to feel when the bolts were snug.

I attached the wedge piece with drywall screws, which my cordless driver/drill sent burrowing into the redwood as far as I needed to countersink them. That's another indispensable tool, especially if you have an extra battery charged up and ready to go.

Once the pieces were connected, I drove a few more screws into each through the bottom rail, flipped the tread over and used a round-over bit in my Sears Craftsman router to smooth the top front and side edges. During use, this router sucks bits up into the collet unless I really bear down with the collet wrench while changing bits—hard enough to bend the tooth on the shaft lock. To switch the bits after such punishment, the whole collet must be removed and the bit forced out with a nail set. Maybe I should just be happy the bit doesn't creep *down* during use.

To install the finished tread, I once again needed to use shims, this time between the tread and the stringers. When I was satisfied that the tread was at the proper height and solidly level, I glued the shims in place with construction adhesive and screwed the cleats into the stringers. I reached back under the tread and screwed the header above to it through the predrilled holes in the header.

When the tread was secured, I cut a riser to fit underneath it. I cut the riser to length, freshened up the face with the power plane and predrilled several holes along the bottom edge, which would later allow me to screw the riser to the back of the tread below it. Then I pried the riser tight against the tread with a flat bar and screwed it to the stringers with the drywall screws, low enough so the screws would be hidden by the tread below. I made and installed the rest of the treads and risers the same way (right photo, facing page).

Any time it was necessary to fit a riser or tread to a rock (left photo, facing page), I transferred the profile of the rock onto it with a contour gauge and gingerly trimmed the ends with a reciprocating saw and a sharp chisel. A contour gauge is made of two metal plates with about 150 tempered-steel wires sandwiched in between. When the gauge is pushed up against an object, the wires slide between the plates until they all touch the object. The contour of the object can then be scribed off the gauge.

Making and installing each tread and riser pair took about two hours. After about a week, I reached the bottom stair and poured a small concrete pad at the base of the stairway.

Finishing touches—Once the stairs were finished, a coat of clear wood finish was brushed on (The Flood Co., P. O. Box 399, Hudson, Ohio 44236-0399). Ron built a nice, tight rock retaining wall against the edge of the stairs to keep dirt and critters from getting under them, and Liz landscaped very nicely around the deck and stairs. I got married and moved away to complete school. Those stairs were an enjoyable challenge and great practice for building an even nicer set of front steps for the Skeltons when I move back to the area.   □

Thor Matteson is a civil-engineering student at California Polytechnic State University in San Luis Obispo, California. Photos by the author.

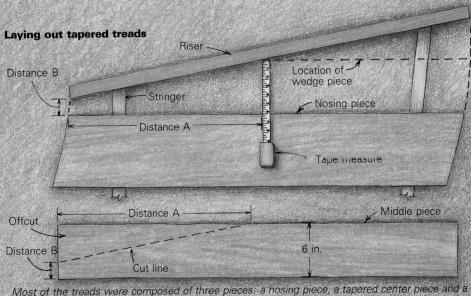

Laying out tapered treads

Most of the treads were composed of three pieces: a nosing piece, a tapered center piece and a wedge piece to fill out the back. Once a nosing piece was cut and positioned on the stringer, the author held his tape measure perpendicular to the rear edge of the piece and marked the point where it measured exactly 6 in. to the riser. Then he measured from the end of the piece to the mark (distance A) and measured the gap between the nosing piece and the riser at the narrow end (distance B). He transferred these measurements to the blank for the middle piece, connected the points with a diagonal line and cut the taper with a worm-drive saw. The technique worked for the small wedge piece, too.

The stair snakes around three boulders en route to the shop, requiring some careful fitting of treads and risers (photo above). The author used a contour gauge to transfer the profile of the stone to the tread and riser stock, then trimmed the stock with a reciprocating saw and a chisel. The stringers are made of 2x12 rough-sawn redwood (photo right), lapped at the joints and secured to wet post anchors and to each other with 3-in. zinc-plated drywall screws. The anchors are made to accommodate 4x stock, so redwood blocks were used to fill out the bottom anchors. Triangular voids in the left-hand stringer were remedied by screwing vertical 2x6 blocks to the stringers to lend extra support to the treads. The risers are screwed to the stringers with drywall screws, low enough so the screws are hidden behind the abutting treads, and they're also screwed to the backs of the treads. Chamfers on the treads prevent the stairs from looking too boxy. A concrete pad will later be poured at the bottom.

Outside Circular Stairway

A handsome addition without fancy joinery yields covered access to two levels

by Tom Law

In the Annapolis, Md., area, tidal creeks and rivers from the Chesapeake Bay produce some steep building lots. My clients owned a ranch-style house on just such a site. To enter the house you had to park in a space along the roadside and walk down steps onto a deck; then to reach the front door you had to walk to the far end of this deck. The owners didn't like this arrangement. The long walk to the front door was inconvenient, and the interior stairway to the basement took up valuable living-room space. An architect neighbor conceived the solution to this problem—an exterior circular stair, located closer to the road, that would provide sheltered access to both the living room and the basement. The finished stairway is shown in the photo below. I got the plans in the form of a freehand sketch. The project involved removing the interior stairs, filling in the floor, building the spiral stairway itself, and cutting a hole in the exposed concrete block wall below the deck to make a new doorway to the basement.

I started by laying out the new entry door about midway down the deck, which still allowed it to open into the living room. The right side of the door was to be the center of the new spiral-stair enclosure. Squaring over to the outside edge of the deck, I dropped a plumb bob to the ground. Using a post-hole digger, I went down about 5 ft. with a 20-in. diameter hole until I hit sandstone. I formed up an octagonal pier of the same dimension 6 in. above grade and poured concrete.

The plan called for a round central post, but I was planning simply to glue and nail the treads in place and felt that an octagonal post would make it easier to shape them to fit. The post was a 10x10 Douglas fir timber, 26 ft. long, which I had to special-order. Once I got it inside my shop I laid it across two sawhorses and chamfered the corners with a Skilsaw. This required tacking a fence on the post and making an initial pass at the greatest depth of cut, and then making an additional rip on the adjacent side of the timber to cut through. To remove the saw marks and even out irregularities I used a belt sander—first across the grain and then lengthwise, trying to get the chamfered surface straight and uniformly wide.

Setting the post was easy, in spite of its length and weight. My son Greg and I slid the post down the hillside steps right up to the pier, and with a rope tied to the top, upended it until we could reach it while standing on the deck. With Greg steadying it from the top, I bear-hugged the bottom end, lifted it onto the pier and seated it over the ½-in. dia. steel pin protruding from the concrete. With the post roped to the deck rail, slightly out of position, I cut out a half-octagonal pocket into the deck band, slid the post into place and nailed it.

On top of the post I cut a pocket for the cross-beam of double 2x6s to carry the roof of the stair enclosure, then beveled off the top of the post for a neater appearance. The roof joists were set on a ledger on the wall and cantilevered over the beam (drawing, facing page). All the sheathing for roof and walls was 2x6 tongue-and-groove fir. Letting the roof deckings run long, I found the center of the post and swung a 3-ft. radius arc. To trim the decking to this line, I used a handsaw for the plumb cuts on the joists, and a reciprocating saw for the curve. I cleaned up the sawn edges by sanding to the scribed line with my belt sander. This new section of roof was finished with flat-seam terne metal held tightly under the soffit, and no changes were made on the existing roof.

Treads and risers—The new basement door was to be directly under the new living-room door, so I marked its location and punched a hole in the basement wall to find the floor level. With a 2-ft. level on top of a straightedge I transferred this elevation to the post. Next I measured the total rise on the post, and calculated the number of risers required to get a riser height between 7 in. and 8 in.—safe, comfortable limits for a single stair rise. I decided on 12 risers. To make sure I had divided correctly and converted the hundredths into sixteenths, I set my dividers and stepped up the post, making sure the top of the last riser would be exactly at the deck level. This required several attempts. An error of only $\frac{1}{16}$ in. can make a difference of $\frac{3}{4}$ in. on twelve steps. When everything was right, I marked and numbered the location of each tread on the post. That done, I went back to the shop.

To lay out a stairway of this complexity, I like to draw full-size plans. On top of two workbenches, I laid a big sheet of clean cardboard

The completed stairway addition to this ranch-style house gives access to the basement without wasting interior space. The new enclosure also offers a covered entry at the deck level much nearer to street parking than was the previous front door.

and with trammel points, I swung a 3-ft. radius arc the width of the stairway. I had a cut-off from the post so I placed it right over the center and traced the octagon onto the cardboard. From the basement landing to the living-room deck, the stairs needed to spiral 180°, so I marked off a semicircle. Using the dividers, I stepped off 11 equal segments, the correct number of treads for 12 risers. Connecting these marks to the centerpoint gave the size and shape of each tread, as shown at right. These radius lines were actually the front face of each riser, and I drew in the line of the 1½-in. nosing on the tread for clarity. Now the pattern was drawn, and all I had to do was transfer the lines on the cardboard to wood. This is where the octagonal post is better than a circular one. You can make straight or angled cuts instead of rounded ones. I did no mortising and used no fancy joinery, but some of the intersections of the treads made excellent connections, almost locking themselves into place on the facets of the octagon. I rounded the outer ends of the treads with the reciprocating saw and belt-sanded to the line where necessary.

Finally, I grooved the underside of each tread to receive the top edge of the riser. Assembly went very well. Starting at the bottom, I put a little glue on the end of the first riser and toenailed it against the post with galvanized common nails, using a drift pin to set the heads below the surface of the wood. Galvanized nails were needed not for their rust resistance but for the coarse shank that would make them resist withdrawal as the treads flexed in use. Placing the first tread over the riser with plenty of glue in the groove, I nailed it securely. After the first two treads were in position and the glue had cured, they were strong enough for me to sit on, and as I added each step, I could walk up and down those already in place. With each new tread I also checked the tread heights against the layout lines.

The next thing was to cut the basement door in the block foundation. I used a masonry blade in my Skilsaw rather than knocking out the concrete block and laying new half-block in a sawtooth pattern. Cutting masonry with an abrasive blade creates clouds of dust and particles. You need an open work area and a good dust mask. This is why I made these cuts before enclosing the stairs. When the wall was cut through and the new door and jamb set, I bridged over to the post for the landing at the bottom of the stairs to form the entry to the basement. This bridge was supported at the block wall on a ledger, and by the post and a temporary prop to the ground on the stair side.

The enclosing curtain wall was also 2x6 fir decking. At the top, each piece was nailed to a joist or to blocking; the bottom was nailed to a plate on the deck, and as I progressed downward, it was nailed to the risers and treads. The boards were held flush at the top and allowed to extend long on the bottom. Again, I applied glue to all joints for strength. The tight joints between the steps and the post allowed no play in the stair, and the vertical fir sheathing connection to the roof added further rigidity and shear strength. The bridge into the basement

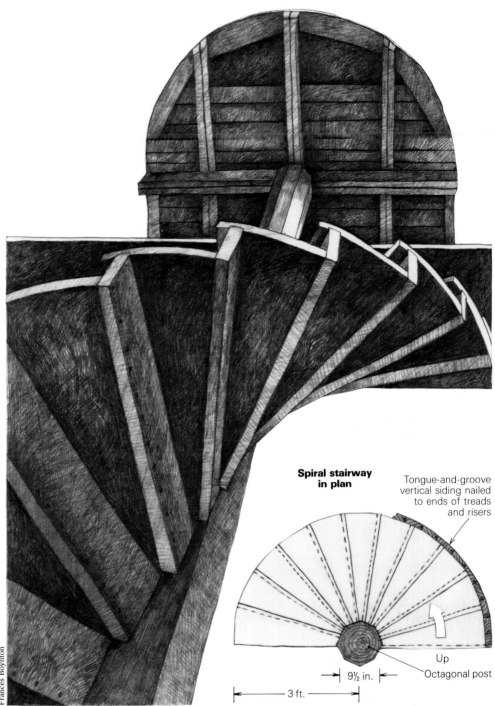

Frances Boynton

Spiral stairway in plan

Tongue-and-groove vertical siding nailed to ends of treads and risers

Up

9½ in.

3 ft.

Octagonal post

The spiraling pattern of treads and risers was laid out full scale in the shop. The bottom of each tread was plowed ½ in. deep to accept the top of the riser; the riser bottom was glued and face-nailed to the back of the tread. When the steps had been glued and nailed to the octagonal post, they were rigid enough to hold the weight of the builder, even without siding to support them. The roof for the stairs is carried by the double 2x6 beam let into the top of the post.

was suspended by the vertical sheathing, which was nailed to the deck joists above; then the temporary prop was kicked out. On the deck level, I cut a doorway into the shell of the stairway on the street side, with a large casement window across from it.

When all the siding was in place with the ends long at the bottom, I marked out the helix, transferring the bottom edge of each riser to the outside of the sheathing. The resulting series of points wrapped around and up the cylinder. Using an ⅛-in. strip of wood, I connected the points to create a line. Starting at the top to get the help of gravity, I just sawed down the length with the reciprocating saw. I took a lot of care with this cut so that it

wouldn't require any further dressing. The stairway was complete.

After cutting out the living-room wall and setting a new jamb and door, the only thing left to do was to remove the old stairs, cover the framed-in opening with oak flooring and finish it like the rest.

The work was done during October and November, and the delightful weather added to my enjoyment of the project. I had a helper for two days when I was mixing concrete and setting the post, but the rest of the time I worked alone. Including repairs to the inside floor and walls, the job took me eighteen days.   □

Tom Law is a builder in Davidsonville, Md.

Building a cantilevered-tread spiral stair

Throughout the Saks house there is a feeling of openness, and we wanted the spiral stair to the bedroom and lookout tower to echo this feeling. We wanted it to be a structure with delicate lines, unencumbered by external support systems and limited to the tones and textures of wood. For me, each aspect of this 6-ft. dia. spiral construction was a challenge, always an adventure, and often a headache. More than once I considered selling my tools and opening a restaurant.

Layout—The central support for the staircase is a 20-ft. long yellow cedar driftwood log, about 1 ft. in dia. at the base. As a structural member of the house, the log was standing when I came on the scene, and it had to be laid out and worked in place. It was neither completely round, nor straight. Consequently, the tread-mortise positions had to be projected inward from a 6-ft. dia. imaginary cylinder symmetrically enclosing the assumed centerline of the log. Since the position of the outside end of each tread was fairly critical, individual tread lengths had to vary by an inch or two depending on the warp of the log. The entire layout for the tread positions had to be completely independent of the log, with the projected mortise positions falling arbitrarily on the log's uneven surface.

The rise of each step is 7.3 in. To climb the distance from the ground-floor landing to the bedroom landing, 13 treads travel an arc of 292° (drawing A)—one tread per 22½°. This let me find the centerline of each tread, and ignore the overlap of the treads, which was an inch on each side.

To get the mortise positions for the treads, I made a flat plywood pattern of the 292½° arc, as shown in drawing B. I drew lines on the pattern dividing it into 13 segments, and made a mark for the centerline of each segment. Then I cut out the center of the pattern so it would fit around the base of the log column, and positioned it on the floor where the first tread would start, minus the overlap. I put some index marks on the pattern and the column for future reference points, then I transferred the centerlines of each tread onto the log using a pencil and a plumb bob. I marked the rise intervals on a story stick and transferred them as top-of-tread lines to the corresponding tread centers already scribed on the pole.

Tread construction—Each tread is completely self supporting. They cantilever from the pole on brackets made from ⅜-in. by 3-in. by 24-in. steel bar stock welded to a ⅜-in. by 3-in. by 6-in. butt plate. A ⅝-in. piece of threaded rod welded to each butt plate extends through the log (drawing D). The butt plates sit in routed mortises ranging in depth from ½ in. to 1½ in., because of the uneven surface of the log.

I laminated the tread blanks from six yellow cedar boards, 1½ in. thick and 3½ in. wide— two halves for each tread were ripped diagonally from a block (drawing C). Two ⅝-in. dia. threaded rods extend through each pair of blanks, capturing the steel flat stock.

After I had the rough treads assembled, I glued on separate knee blocks at the base of the steel butt plates. I rough-shaped the treads with a drawknife, and fine-tuned them with a scorp, spokeshave, and where possible, with the front roller of a belt sander. I filled the gap left between the tread halves with a ⅜-in. square cherry strip.

As it turned out, the biggest flaw in the entire process was the applied-knee idea. The knees were difficult to position, their feathered gluelines often showed, and the color and grain were hard to match. It would have been simpler and probably less expensive in the long run to begin with thicker blanks of cedar and cut out the basic shape on a big bandsaw.

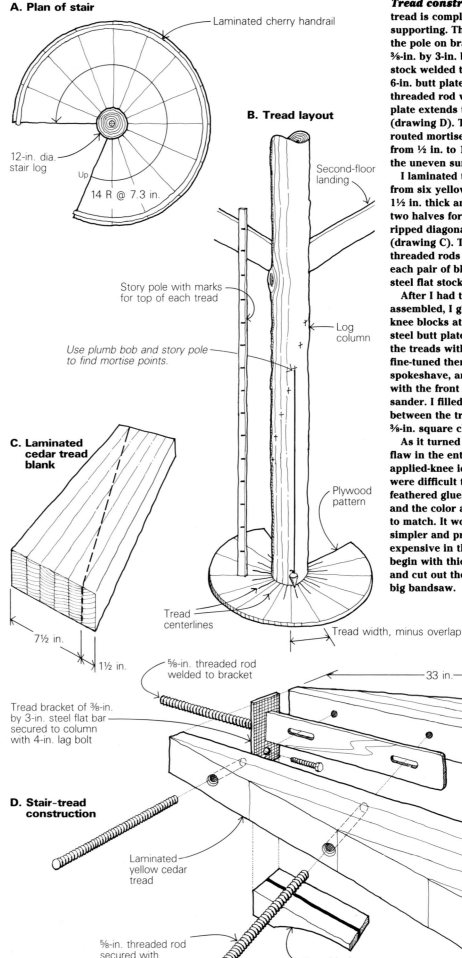

A. Plan of stair

Laminated cherry handrail

12-in. dia. stair log

Up

14 R @ 7.3 in.

B. Tread layout

Second-floor landing

Story pole with marks for top of each tread

Use plumb bob and story pole to find mortise points.

Log column

Plywood pattern

Tread centerlines

Tread width, minus overlap

C. Laminated cedar tread blank

7½ in.

1½ in.

⅝-in. threaded rod welded to bracket

33 in.

Tread bracket of ⅜-in. by 3-in. steel flat bar secured to column with 4-in. lag bolt

D. Stair-tread construction

Laminated yellow cedar tread

⅝-in. threaded rod secured with countersunk nuts and washers

Knee block

Mortising the column—For each tread I had to cut mortises for the tread butts, drill the bolt holes through the column, and counterbore for nuts and oversized washers.

I made multiple passes with a ½-in. straight-flute bit in a heavy-duty router to cut the mortises. To guide the router, I used a jig that could be positioned and secured to the column with clamps (top photo). Accuracy here was crucial because the flat bottom of the mortise formed the seat for the butt block of the tread bracket. The slightest inaccuracy in any plane would translate out to the end of the tread, putting it hopelessly out of position.

I made the jig from a piece of ½-in. plywood, 12 in. by 18 in., with 1x2 fences for the four sides of the mortise. For clamp jaws, I used 18-in. long pieces of 2x2 at the top and bottom of the jig, and opposite the column. They were linked by threaded rods. To bring the jig into plumb, I checked the bed on which the router moved for plumb in both planes, and adjusted it with wooden wedges.

Tangential orientation of the jig was the most crucial, and the most difficult. The jig's flat surface had to be perpendicular to the final centerline of the tread, and this angle had to be set from a point as far from the pole as possible in order to minimize error. To do this I added a 14-in. plywood tongue extending horizontally from the lower edge of the jig, with the tread centerline marked on it. I adjusted the jig with wedges between its base and the column until the centerline on the tongue was plumb with the corresponding center marked on the plywood pattern on the floor.

Mortises routed, I drilled bolt holes using a guide block that fit into the mortises. It both indexed the hole position and trued the ⅞-in. auger bit so that I could drill a hole perpendicular with the mortise bottom. I did my counterbores with an expansion bit. To center the spur on the bit, I tapped square wood plugs into the last 2 in. of the bolt holes, and then drilled them out during the counterboring. Leftover plug ends came out easily using a length of dowel as a drift pin.

The test—At this point the finished treads were bolted in, and the staircase was ready for a test climb. Since the railing had not yet been built, the treads were supported only by the center pole. Up to this point the

To cut the tread mortises in the column, Grunewald used this jig to guide the router. For this cut near the landing, he removed the top pair of 2x2 clamping jaws, and nailed the jig to the log column.

By clamping gusseted brackets to the stair treads, Grunewald was able to use the stair as the form for the laminated handrail. The edge of each bracket was positioned to correspond with the line of the handrail, and their upright legs served as clamping surfaces during glue-up.

engineering had been theoretical. For a practical test, I asked an unbiased (and well-insured) friend to try out the stairs and to report on three things: bounce, wiggle and ease of ascent. After a couple of trips he said that the stairs were easy enough to climb and didn't deflect much vertically, but that they wiggled unnervingly.

It was clear that the treads had to be tied together to stiffen the whole structure. The obvious solution was to use the balusters, fastening them into one tread and against the face of the next tread up. But this couldn't be done until the handrail was built.

Handrail—I laminated the rail out of nine layers of ³⁄₁₆-in. by 3-in. cherry strips, with the joints staggered at 2-ft. intervals. The staircase became the form, with gusseted angle brackets on the treads providing clamping points and forming the helical curve (bottom photo).

I had originally thought to glue up all nine layers at once. But since the weather was hot and I was workng alone I elected to do two separate gluings, one of four layers, the next of five, hoping to avoid some of the panic that glue-laminating can cause. That turned out to be a wise decision.

After it cured, we plucked the handrail from its form and planed it on the Hitachi jointer/planer. To do this, I laid the machine on its side on the porch, and fed the rail through like a huge corkscrew to a helper stationed downhill. Once it was planed, I put it back in place and marked the baluster positions. I drilled these using an angled guide block that clamped to the outside face of the rail. I cut the balusters from 1-in. square cherry blanks. Rather than turning them on the lathe, I found it quicker and easier to do four passes over a table-mounted router with a ½-in. roundover bit. I split the bottom of each baluster for a blind wedge, and filed a flat spot to seat firmly with the edge of the adjacent tread.

Each baluster was then wedged and pinned into the top face of a tread and screwed to the front edge of the tread above, effectively tying the staircase together and eliminating almost all of the movement. The predrilled rail was then tapped onto the upper ends of the balusters, and except for the many hours of sanding with 150-grit paper and applying tung oil, the job was finished. □

John Grunewald lives on Hornby Island, B. C.

Backyard Spiral

A stair of redwood and steel built to endure the elements

by Richard Norgard

The spiral stair links the deck, which is attached to the main floor of the house, to the backyard. The stair consists of a column made of steel pipe, with steel brackets welded to the column, and redwood treads bolted to the brackets. Steel balusters support a glue-laminated redwood handrail. The stair rests on a 16-cu.-ft. chunk of concrete.

Before I became a self-employed carpenter, I used to design and build scenery for the professional and educational theater. That's where I learned that spiral stairs date back to the Middle Ages and that, in those days, they typically coiled counterclockwise from top to bottom so that a right-handed defender with a sword could easily fend off a right-handed attacker coming up the stairs.

Working in the theater, I built several small, temporary spiral stairs, but I never constructed a permanent spiral stair. When Dan and Patsy Mote hired me to design and build a deck and a spiral stair onto the back of their house, I finally got the chance. The stair would not only be permanent, but it would also have to be built to beat northern California's weather.

The overall plan—The Motes' house, which is located in the Berkeley hills, is designed to accommodate large groups of people. The focal point for entertainment is the second-floor living room, which faces San Francisco Bay and overlooks a spacious backyard. The Motes wished to add a deck off the living room so they could entertain their guests both indoors and outdoors. But equally important, they wanted a new set of stairs to connect the deck with the backyard below.

After a series of discussions, we settled on a design that would go relatively easy on their pocketbooks and that would also accommodate future additions. The design called for a 200-sq.-ft. redwood deck to be built off the living room, with a spiral stair linking the deck to an existing deck in the backyard (photo above). Access from the house to the new deck would be via a pair of matching French doors that would replace an existing picture window in the living room. Instead of the usual piers, the deck would bear on a stepped concrete foundation. This would someday allow a sun room to be added beneath the deck. In the mean-time, the foundation would contain a work area paved with brick embedded in sand.

Carpenter Tim White, assistant James Moore and I built the stepped foundation and deck with little trouble. We left the crux of the job, the spiral stair, for last.

Designing the stair—The Uniform Building Code only marginally covers spiral stairs. The code simply requires that treads measure 26 in. from the outside of the supporting column to the inside of the handrail, that the run is at least 7½-in. at a point 12 in. from the narrow end of a tread and that the rise of the treads be no greater than 9½-in. (as opposed to the usual 8 in. for straight residential stairs). Regardless of the code, though, what's legal in one area may not be legal in another. Before I built my stair, I discussed in detail its materials and construction with the local building inspector and plan checker.

The final stair plan called for a column made of 3½ in. by ¼-in. thick steel pipe, with steel brackets welded to the column and redwood treads bolted to the brackets (middle drawing facing page). Square steel balusters would support a glue-laminated redwood handrail. Though my clients didn't plan to do any sword-fighting on their stair, a counterclockwise spiral fit best on the site.

The whole stair would weigh over 1,200 lb., enough to require a substantial footing to support it. We designed a massive reinforced-concrete slab that would both support the weight of the stair and anchor it to the hillside (bottom drawing facing page). The 16-cu.-ft. slab is shaped like an inverted truncated pyramid. It's reinforced with two horizontal triangles of ½-in. rebar, one above the other, with three vertical lengths of rebar tied to the points.

Concrete and steel—Because the stair landed on an existing deck, we had to remove the old decking before we could form and pour the slab. During the pour, we centered a ¾-in. thick by 12-in. by 12-in. steel plate on top of the slab and tied it to the concrete with four ½-in. anchor bolts. Once the concrete set, we positioned the steel column on the slab, plumbed the column and welded it to the plate. After the treads were installed, we drove a cylindrical piece of redwood into the top of the column and screwed it to the column with galvanized drywall screws (we call them "deck" screws). That allowed us to screw the deck handrail to the top of the column. The column would also be secured to a deck post below the joists with a pipe collar and a length of ½-in. threaded rod. A turnbuckle in the threaded rod would allow future adjustments. Once the column was welded in place, we reinstalled the decking on the existing lower-level deck.

The tread brackets were made by a local welding shop. Each bracket consists of a 3½-in. wide by 22¾-in. long piece of flat steel welded to a 7-in. length of steel tubing (top drawing right). The steel tubing was just the right size to allow us to slip the treads over the column later and weld them in place. Three 1-in. holes punched through each bracket with a punching machine would accommodate the bolts for attaching the wooden tread pieces to the brackets. To prevent rust, all steel in the stair, including the column, was finished with two coats of steel primer and three coats of heavy-duty exterior lacquer.

Treading softly—The tread pieces were cut out of select redwood 2x4s, except for the pie-shaped pieces on the outboard ends, which were cut out of 4x4s. We used redwood because it has excellent water-resistant and structural properties, it doesn't attract fungus or bugs and it's relatively inexpensive in northern California. By drawing a full-scale tread pattern first, we were able to determine the exact length and shape of each piece. We cut all the pieces on a 12-in. radial-arm saw and drilled the bolt holes using a drill press. We used a machinist's vise to position the pieces accurately.

Because this was an exterior stair, it was important that we provide gaps between the 2x4s of the treads for drainage. The galvanized washers commonly found in hardware stores are of an inconsistent thickness, so I used stainless-steel washers instead. Besides being of a uniform thickness (three of them created a gap of ⁵⁄₁₆-in.), they wouldn't rust or stain the wood. I found mine at Bowlin Equipment Co., 1107 10th St., Berkeley, Calif. 94710.

Once all the stair pieces were cut and drilled, we bolted them to the brackets with ½-in. galvanized carriage bolts, counterboring the bolt heads and nuts. Once the bolts were snug, we cut the ends flush using a reciprocating saw. Finally, we fixed a 2x4 cap to the end of each tread with galvanized drywall screws and sanded the tread tops with 36-grit sandpaper in a belt sander. This flattened the treads and created a slightly rough surface for better traction.

After sanding, the treads were ready for installation. As is often true with spiral stairs, the nosings (the overlaps of adjacent treads) on this stair taper. In this case, there's a 2-in. nosing at the column and no nosing at all at the outside edge. At first, the building department had a problem with this, but once they checked their catalogs and saw other stairs built this way, they acquiesced.

Using steel tubing as a collar worked out great. First we slipped all the treads over the

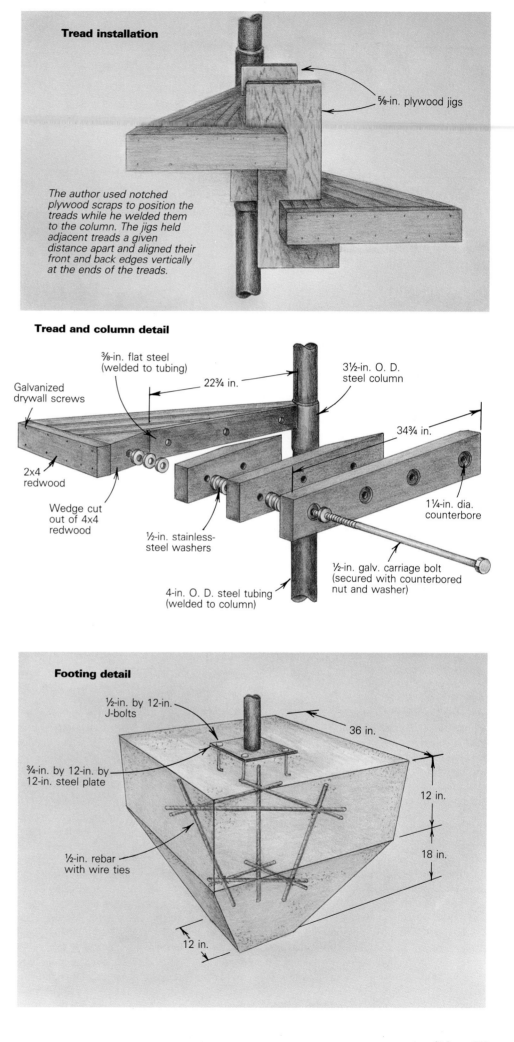

Tread installation

The author used notched plywood scraps to position the treads while he welded them to the column. The jigs held adjacent treads a given distance apart and aligned their front and back edges vertically at the ends of the treads.

⁵⁄₈-in. plywood jigs

Tread and column detail

⅜-in. flat steel (welded to tubing)

Galvanized drywall screws

22¾ in.

3½-in. O. D. steel column

34¾ in.

2x4 redwood

Wedge cut out of 4x4 redwood

½-in. stainless-steel washers

1¼-in. dia. counterbore

½-in. galv. carriage bolt (secured with counterbored nut and washer)

4-in. O. D. steel tubing (welded to column)

Footing detail

½-in. by 12-in. J-bolts

¾-in. by 12-in. by 12-in. steel plate

36 in.

12 in.

½-in. rebar with wire ties

18 in.

12 in.

The balusters (photo above) are made of square steel tubing with steel mending plates welded to the bottoms and a pair of angle-iron brackets welded to each top. The mending plates are lag-screwed to the backs of the treads and the handrail is attached to the top brackets with galvanized drywall screws. A lag screw anchors each baluster to an upper tread and eliminates vertical movement of the treads. The handrail is composed of six plies of redwood (photo left) which were glue-laminated on site. Aluminum connector bolts and cap nuts add a decorative touch and also help ensure against delamination. The photo shows the joint between the curved railing and a short horizontal return at the top of the stair. The rail is protected with three coats of spar varnish. *Photos by author.*

column and secured the pipe collar at the top of the column to the deck post. Then my associates lifted the top tread into position and held it there while I welded the collar to the column. The rest of the treads went up easily with the aid of a pair of jigs. The jigs were simply plywood scraps notched to hook over the nosing of a tread and to hold the adjacent tread a given distance below it (top drawing previous page). The jigs also served to align the front and back edges of adjacent treads vertically (at the end of the treads). Using a 2-ft. level to ensure that the treads were level, I welded the rest of the treads to the column. The final rise of each tread worked out to be 8⅝-in., well within code.

Bending the rail—The construction of the glue-laminated handrail both excited and concerned me (as glue-ups usually do). Because this railing was subject to both rain and sun, the materials, including the wood, glue and finish, had to be weather-resistant. For the sake of appearance, I also wanted the wood to be of a mixed color and free from knots. With the help of Jeff Hogan, owner of Ashby Lumber in Berkeley, I found what I

was looking for. Ashby Lumber carried pre-packaged "A" grade redwood boards which ranged in color from blond to reddish-brown. The boards were ⅜ in. thick by 3¼ in. wide and were surfaced on one side, and they came in random lengths (most of them measured between 8 ft. and 10 ft. long). I ripped the boards to 2⅜ in. wide on my table saw and thickness-planed them down to slightly more than ¼ in. thick. After I carefully arranged the boards according to color, they were ready for glue-up.

The finished handrails were to be 1⅝ in. wide (six plies) by 2¼ in. high. I used two-part resorcinol glue, which is waterproof, and staggered the butt joints between boards a minimum of 2 ft. For the glue-up forms, I temporarily attached to each tread a plywood gusset with a short length of 2x2 screwed to it. The gussets were attached to the treads with screws. Later, the screw holes would be covered by the balusters. Because the railing would have to be clamped every 4 in. or so, I collected every C-clamp I could find and then made the rest out of plywood scraps and carriage bolts with wing-nuts. The glue-up took three days to complete because I could man-

age only two plies per session without the glue setting up and without running into problems staggering the joints. Though the label on the can of glue recommended leaving the clamps on for 12 hours, I left them on for 24 hours.

Once the glue-up was finished, I removed the 27-ft. long handrail from the stair and planed the top and bottom smooth with a portable electric plane. I then eased the edges with my router chucked with a ¼-in. roundover bit.

Building the balustrade—The balusters were cut out of ⅛-in. thick by 1-in. square steel tubing (top photo left). I welded a ⅛-in. by 1-in. by 5-in. steel mending plate to the bottom of each baluster, extending the plate 3 in. below the end of the tubing. I cut the tops of the balusters at an angle to match the slope of the railing and welded a pair of pre-drilled 3-in. angle-iron brackets to the top of each baluster. I finished up by bending the brackets to the proper angle and rounding their corners with a grinder.

To install the balusters, I rested them on the treads and screwed the mending plates to the back edges of the treads with ⁵⁄₁₆-in. by 3½-in. lag screws fitted with lock washers. Each baluster touched the nosing of the tread above and was screwed to it with the same lag screws and lock washers. This totally eliminated vertical movement between treads.

After bolting the balusters to the treads, I positioned the handrail on top of the balusters and scribed the shape of the angle-irons on the underside of the railing. Then I removed the railing and by eye routed out mortises for the brackets using a router and a ½-in. mortising bit; I wanted the brackets to be flush with the bottom of the railing. Because the corners of the angle iron were rounded, there was no need to chisel out the corners of the mortises. Once the routing was complete, I repositioned the railing and secured it to the balusters with 1⅝-in. long galvanized drywall screws. Up top, the transition to a short horizontal railing section gave us the chance to display some tidy joinery (bottom photo left).

As a decorative touch and an extra precaution against delamination, I drilled ¼-in. holes through the railing and installed aluminum joint-connector bolts and cap nuts on 6-in. centers. These bolts can be adjusted from 1¼ in. to 2 in. in length and come in brass, aluminum or stainless steel. The bolts added a distinctive nautical appearance to the stair. I got mine at a local boat shop.

Finish-sanding of the railing went quickly. As the final step, I brushed on three coats of high-gloss spar varnish. This not only assured retention of the color in the wood, but also eliminated any possibility of splintering. ☐

Richard Norgard is a free-lance carpenter in Fort Bragg, California.

Quadrant stairs. The seven stringers, fanned out from an axis, and the bandsawn treads on this exterior staircase gave the author and his partner an opportunity to depart from the common world of rectilinear carpentry.

Fantail Deck Stairs

Radiating stringers and concentric treads result from simple geometry

by José L. Floresca

I started working with curves about 10 years ago. I learned from a master carpenter, Tom Pratt, who taught me how to build curved forms for concrete pools. What made this different from rectilinear construction was that we used plywood for the top and bottom plates. To follow the different radii for the curved surfaces of the pool, we cut the plates with a jigsaw or a bandsaw. A similar technique worked when I built a circular soffit in my kitchen. But my appetite for challenging curved projects hadn't been really nourished since Pratt moved out of town in 1987.

A new challenge presented itself when my partner, Steve Cassella, and I saw the plans for a remodel/deck addition designed by Dwayne Kohler that included a set of exterior stairs that fanned out in a quadrant, or quarter circle. Questions arose about the stair's construction: How many stringers are necessary to support the treads? How should the stringers be supported? What kind of calculations would be necessary to build the stairway?

After we got the job, and before we could build the quadrant stairs, we first enlarged the home's existing upper deck and built a new lower deck. That was the easy part of the job.

Calculations and cogitations—The plans called for a straight set of stringers to run from the upper deck down to a landing. From the landing a second set of stringers would fan out in a quadrant (photo previous page). The total rise of the stairway—from upper deck to lower—is 93 in. We divided the height into 12 risers. More risers would have meant more treads, resulting in longer stringers. And longer stringers would have taken up more space on the lower deck. The rise (93 in.) divided by the number of risers (12) gave us an individual riser height of 7¾ in. We cut 10-in. treads because 10 in. is a good tread width for three 2x4s with ⅛-in. drainage space between the boards and a ¾-in. nosing.

After we determined the length of our stringers, we framed the landing and installed the straight-run stringers to the upper deck. We could now concentrate on the quadrant stairs.

The calculations for cutting the stringers of the quadrant stairs are the same as those for the straight run above. The big question was how many stringers did we need to support the treads? We decided that the span between the stringers at their widest point (the outside edge of the bottom riser) should not exceed 20 in. because the grain of the treads is not consistently perpendicular to all of the stringers they rest upon.

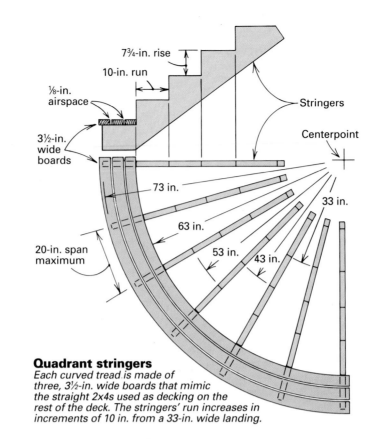

Quadrant stringers
Each curved tread is made of three, 3½-in. wide boards that mimic the straight 2x4s used as decking on the rest of the deck. The stringers' run increases in increments of 10 in. from a 33-in. wide landing.

The landing is 33 in. wide. We used this dimension as the radius for the top risers of the quadrant-stair stringers. Because our treads are 10 in. wide, the radii for the other risers are larger in 10-in. increments (the width of each tread); i. e., 43 in., 53 in., 63 in. and 73 in. (drawing above).

To calculate the number of stringers for the bottom step, we first had to calculate the circumference of the circle formed by the radius of the bottom step. Knowing that $2\pi r$ = circumference (r is the radius), we plugged in the numbers: 2 x 3.1416 x 73 = 458.6736 in. The stairs occupy a quarter circle, so we divided the circumference by four. This figure was divided by the desired stringer span, 20 in., which gave us the number of 20-in. spaces in the quarter circle. Rounding our quotient to the nearest whole number gave us six spaces, so we'd need seven stringers.

A curved brace for the stringers—Installing the stringers posed yet another challenge.

A curved brace. All the stringers are supported by a single, curved brace, which was notched to accommodate each stringer. A 2x10 leg runs between the deck joist and the brace.

How should we support the tops of the stringers? Rather than support each stringer individually, I visualized a single, curved brace that each stringer would bear against.

I cut the brace from a 2x12. The curve of the brace is a segment of a circle. Its radius was determined by finding a centerpoint that was perpendicular to the center stringer and equidistant to the backs of all the other stringers, roughly 3 in. from their tops. We added 1 in. to the circle's radius to allow the brace to be notched around each stringer for lateral support. The brace was attached to the framing in such a way that its face is perpendicular to the back of the center stringer (photo below). A 2x10 leg supports the curved brace and was nailed to the top of one of the deck's joists. The top of the leg was cut at a compound angle and nailed to the curved brace (photo below). With both pieces nailed in place, we then notched the curved brace with a jigsaw to accommodate the stringers (top photo, facing page).

The bottoms of the stringers sit on the lower deck and are toenailed in place. To provide nailing for the landing's deck boards, we added blocking between the stringers about 3 in. behind the top risers.

Determining tread length—To calculate the length of the treads along their circumference, we multiplied the radius of the lowest riser by two to get its diameter. We then multiplied this diameter by π to obtain the circumference. Because the arc of the stairway is a quarter circle, we divided the circumference by four.

For each of the four treads, it was necessary to cut three rows of boards (to simulate 2x4s): the inner board, which butts up against the riser above; the middle board; and the outer board whose nose hangs over the riser below. Each board of each tread has both an inside and outside radius. The inside radius of the inner board is the same as the radius of the riser above—33 in., 43 in., 53 in., etc. The outside radius of each board is always 3½ in. (the width of a 2x4) greater than the inside radius. The inside radius of the next concentric tread board is greater by ⅛ in. (airspace) than the outside radius of the previous tread.

To calculate the lumber necessary for the treads, we used the same formula as that used to determine tread length. We calculated for each of the 12 tread boards plus one for the quarter-circle edge on the landing. To our total lengths we added an additional 10% as a safety factor. We added it up and purchased tight-knot cedar 2x10s in this linear quantity. Although this figure was derived as a length along a curve, it allotted for additional material needed so that we could stagger joints on the stringers.

Marking and cutting the treads—Accuracy was essential, so we laid out the curved treads on the level floor of our cabinet shop. Cassella made a marking device out of a 1x2, an adhesive-backed tape measure and a set of trammel points. One point was set at zero. The other was adjusted to the lengths of the curves determined when we calculated the tread lengths (drawing p. 112). When marking the curves on the 2x10s, we avoided using checked ends, and we placed knots so as not to create holes or cracks.

We first cut the curves with a jigsaw, which was inadequate because the cuts were not consistently square. The bandsaw cut the curves, but it became a two-man job because of the long lengths involved. We left the ends of the boards long and cut them to length during installation. We routed a ¼-in. radius on all visible edges.

Installing the treads was fairly straightforward. We predrilled the nail holes where necessary so that we wouldn't split the ends, and we also staggered the joints (photo below).

Curved work has a natural character to it. It adds life to the otherwise rectangular dimensions of most carpentry. Enter some geometry, a bandsaw and, voila, quadrant stairs. □

José L. Floresca is a carpenter in Seattle, Wash. Steve Cassella coauthored this article. Photos by José L. Floresca except where noted.

Stringers in place. The calculations for determining rise and run for the stringers for the quadrant staircase are identical to ones used on conventional, parallel stairs. The stringers differ from the ordinary in that they radiate from a common point.

Installing the treads. Three 3½-in. wide curved boards make up each tread. The curves were laid out using a marking device made out of a 1x2, an adhesive-backed tape measure and a set of trammel points. After layout, the authors cut the boards on a bandsaw.

Building an Exterior Newel Post

Redwood boards and custom moldings decorate a pressure-treated post and anchor the front-porch steps

by Peter Carlson

I could hardly believe my good fortune when I landed a job at Preservation Park. This development in Oakland, California, is a collection of historically significant houses that were rescued from the wrecking ball. The dozen or so houses that make up the project were neglected, run-down and in the way of other projects. But instead of carting them off to the landfill, the local redevelopment agency had the foresight to move the houses to a new neighborhood where they could be rebuilt into offices that honor Oakland's diverse architectural heritage.

The houses, which ranged in style from Tudor to Victorian, needed rebuilding. I was to rebuild the porches and the stairs, which had been lost during the moves. The newel posts I built for Trobridge House (photo below) are good examples of the work I did at Preservation Park. And while Trobridge House is a Victorian of the Italianate style, the methods I used to build the newels certainly could be adapted to other styles.

Custom-milled moldings—The Italianate style enjoyed the height of its popularity in the 1870s.

The style drew heavily on massive, classical masonry motifs for inspiration, translated into wood by the ingenious millworking machines of the Industrial Revolution. The ubiquity of mills made highly finished materials readily available to the contractor for jobs both big and small.

Long gone is the vast selection of off-the-shelf Victorian house parts. But here in the San Francisco Bay area, there are still a few mills that can duplicate the old trim. Guided by bits of paint-encrusted molding from Trobridge House, the venerable El Cerrito Mill and Lumberyard in the nearby town of the same name ground shaper knives to mill crown, base, panel moldings and handrails for this job (left drawings, facing page).

Tied to the ground—As the primary anchorage for the handrail, the newel post must be firmly connected to the landing. The most typical method (and the one used on this job) is to build the newel around a pressure-treated post that is bolted to a steel post base. Set in concrete, the base keeps the bottom of the post from getting wet by elevating it slightly above the ground. As shown in the right drawing on the facing page, the base should be oriented so that the bolts are perpendicular to the line of the handrail with the prongs helping to stabilize the newel from the side. The steel post base requires careful layout because the location of the newel must be decided before the concrete is placed.

Start by drawing—The newel and the handrail are among the most visible of finish details, and the materials to make them are often complicated and expensive. Even if the working drawings look to be accurate and complete, I do a full-size elevation and section drawing of a newel post before I start cutting materials. In their reduced size, working drawings have a way of obscuring problems that show up at full scale.

The newel is square in plan and has panels framed by panel moldings, rails and stiles (the horizontal and vertical members, respectively, of a frame). The heart of the newel is a simple box, the sides of which ultimately become the faces of the panels. The dimensions of the panels govern all other parts, so I began by drawing them, which determined the size of the box. Once I made my drawing, I could take direct measurements for all the parts.

The elevation drawing also allows me to study the intersection of the rail with the newel. With it

Italianate Victorian stair. **Massive shapes inspired by classical stonework distinguish the Italianate style. The newel posts, which support the handrail at the first tread, conceal pressure-treated posts anchored to the concrete landing. Because the house is now a commercial office, auxiliary handrails were required to meet contemporary codes.**

Start with a box. A simple box is the core of this multilayered newel post. A piece of solid wood at the end helps keep the pieces square during assembly. Here the author aligns a stile assembly before nailing it to the box.

Clamp, then nail. Rails and stiles are held in position by pipe clamps while they are nailed. The rails should be about 1/32 in. long to ensure no gaps between the rails and the stiles.

The trim carpenter's friend. The rabbeted edge of a panel molding allows the trim pieces to be loose-fitting while concealing any gaps between the trim and the frame.

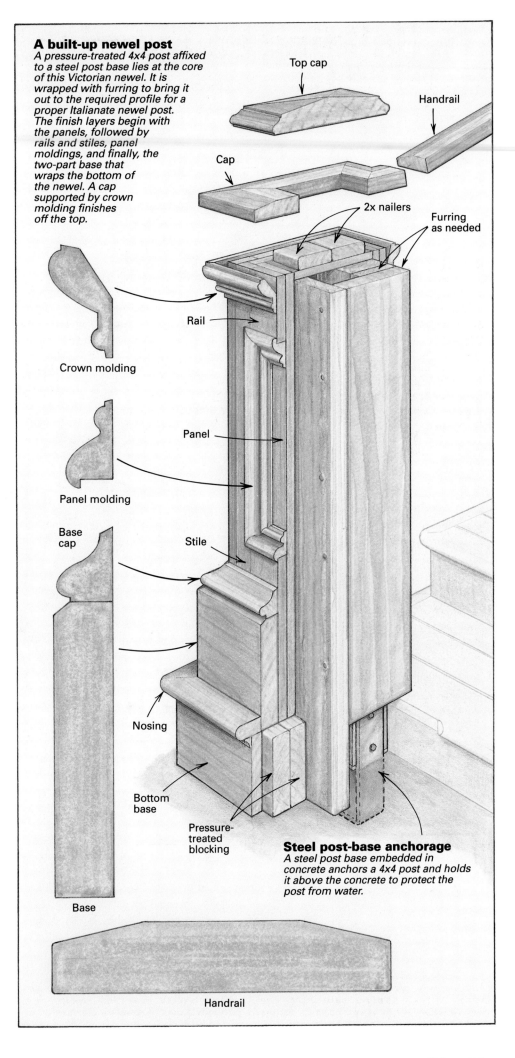

A built-up newel post

A pressure-treated 4x4 post affixed to a steel post base lies at the core of this Victorian newel. It is wrapped with furring to bring it out to the required profile for a proper Italianate newel post. The finish layers begin with the panels, followed by rails and stiles, panel moldings, and finally, the two-part base that wraps the bottom of the newel. A cap supported by crown molding finishes off the top.

Top cap

Handrail

Cap

2x nailers

Furring as needed

Crown molding

Rail

Panel

Panel molding

Base cap

Stile

Nosing

Bottom base

Pressure-treated blocking

Steel post-base anchorage

A steel post base embedded in concrete anchors a 4x4 post and holds it above the concrete to protect the post from water.

Base

Handrail

Base wrap. A mitered 2x frame forms the base of the newel post. Before attaching the frame to the newel, the author used a block plane to knock down any high spots at the butt joints between rails and stiles.

Stiles

Rail

Panel

Base

Far out the post. Pressure-treated lumber nailed to the post stabilizes the newel and provides backing for the newel. The 2x nailers atop the newel are for attaching the cap.

Headed home. The author slips the nearly completed newel over the built-up post. The post is about ¼ in. smaller than the inside of the newel, which allows some room for adjustment.

I can determine the post height, the angle of the handrail intersection and the true proportion of panels, crown and other trim. Superintendents often balk at the time I spend making these drawings. I remind them of the old chestnut, "The only time thinking is seen is when it is not there."

Decorating the basic box—I assemble the box around squares of solid wood or plywood (top photo, p. 115). Typically the lower square is left loose and used only for layout, and the top is permanently installed as backing for the trim pieces that cap the newel.

The stiles are affixed to the corners of the box core—like corner boards on a clapboard-sided house. They are long and narrow and run the full height of the newel. To develop a solid corner, I avoid miters—they always open up over time when they're outdoors. Accordingly, one stile must be ¾ in. (the thickness of the material) smaller than its mate to maintain symmetry. Here's a simple way to rip both pieces with just one table-saw setting: Set the fence and rip the wider stile, then place a piece of the stock you are using against the fence when you cut the second board. This will quickly and accurately give you the setting for the narrower stile.

Before attaching them to the box, I preassemble the pairs of stiles, using nails and an exterior-grade glue. I use Tightbond II, the waterproof yellow glue. I think it's adequate given the fact that the surfaces will be painted. When I start nailing the pieces to the box, I'm careful to place fasteners, wherever possible, in spots that will be concealed by subsequent layers of trim. And if the design calls for edge routing, I keep fasteners out of the line of the router's cut.

By the way, I've become a believer in back-priming (even though the superintendent claims we're doing the painter's job). Before I attach the various layers of material to the boxes, I squirt their backsides with a water-based acrylic sealer. This step helps keep the tannins in the redwood from leaching into the final paint job and mini-

mizes swelling and shrinking that result from changing weather conditions.

I cut the rails about 1/32 in. long to make sure that the assembled frames weren't held apart by the underlying box (middle photo, p. 115). Once I had the rails and the stiles affixed to the boxes, I applied the panel molding. A true panel mold is designed to be the carpenter's friend. Along one edge it has a narrow rabbet the depth of the stiles and the rails. The lip of the rabbet covers the perimeter of the panel (bottom photo, p. 115). The rabbet allows the carpenter to fashion accurate miters quickly by slightly undersizing the molding. The rabbet lip hides gaps between the molding and the rails or stiles. You just cut the pieces to fit a little loosely, assemble them into a picture frame, pop the frame into place and nail it.

I attached the top half of the baseboards next (top left photo, facing page). On the finished newel, the nosing of the bottom tread wraps around the newel (photo, p. 114), becoming a design element in its own right. I installed the nosing and the bottom half of the base after the newels were installed. That way I could scribe the bottom base to fit the concrete landing.

Installing post and cap—I added pressure-treated furring to the post for solid backing (bottom left photo, facing page). Each furring piece engages the concrete or the first tread, adding stability to the assembly. I left about ¼-in. gap between the furring and the inside of the box to allow for adjustment. Then I slid the newel over the post (right photo, facing page). If the layout of the post base is off, you can make final adjustments at this point so that the newel will be plumb and square to the handrail. The post was shimmed solid and fastened with 3-in. long galvanized screws placed where they would be concealed by layers of trim (left photo, below).

There are two basic designs for newel caps used on the houses in Preservation Park: unconnected handrail and connected handrail. The first is the easy one because the handrail simply

dies into the side of the newel below the cap. The cap protects the interior of the newel. In addition to protecting the newel's interior, the second cap design has to include a mitered transition for the handrail (middle photo, below).

In the heyday of Victorian construction in Oakland, the simple cap seems to have been more common. I've found that working drawings for restoration projects often complicate the originals. This job, of course, called for the more complicated of the two. My full-size drawings were invaluable in figuring out the angle of the miter required to meet the rail.

I made the cap from pieces of the redwood handrail that I ripped in half. The miters around the newel cap get all the weather, so they must be crafted carefully. The stock should be kept dry and allowed to acclimate before assembly. I backprimed each piece, used plenty of glue in the joints and cross-nailed them with pneumatic fasteners. I bedded the cap in a layer of caulk around the perimeter of the newel.

A frame of crown molding supports the bottoms of the cap pieces (right photo, below). Like the rails, I cut the crown pieces a bit long to avoid fussing over the joints. I tacked finishing nails to the rails and the bottom of the cap to hold the crown moldings as I tested them for fit. For this kind of dry fitting, miter clamps are very useful. I use the West German ones made by Ulmia (dist. in the U. S. by Robert Larson, 33 Dorman Ave., San Francisco, Calif. 94124; 415-821-1021). Once I dry-fit the crown moldings, I fasten them with the pneumatic fasteners. The clamps are equally indispensable for the base cap.

A top cap made from a single piece of stock completes the top of this newel. Because it has some end grain exposed to the weather, I applied several coats of primer to it before bedding the top cap in a thick bead of caulk. □

Peter Carlson lives in Oakland, Calif., and is a carpenter and member of Carpenters Union Local 713. Photos by Charles Miller except where noted.

Screwed down tight. Trios of 3-in. long galvanized screws secure the newel to the post at top and bottom. The screws are driven in places where they will be concealed with trim.

Double-duty cap. Most newel caps simply keep the weather out of the newel. This one, however, includes a beveled piece of handrail for intersecting the handrail of the balustrade.

Clamping crown. A band of crown molding flares out the base of the cap while supporting it. Here the author dry-fits the pieces using miter clamps to hold the corners together.

A Dry-Laid Stone Stairway

This stonemason uses a crushed-rock-and-gravel method to build well-drained, tightly laid steps

by Dick Belair

Stonemasonry is physically demanding and challenging work. Fortunately, the rewards for breaking your back are quite substantial. The long-lasting beauty of a well-built stone project is as pleasing a sight for a stonemason as it is for its user.

Probably the most challenging and handsome kind of stone project I build is the dry-laid stone stairway, where tight joints are essential for both beauty and strength. Dry-laid masonry work means that the stones are doing all the work—there's no mortar anywhere.

Designing a stone stairway—Locating the stair is the first design consideration. Plan access around existing circulation patterns and look for any existing paths down the hill. Another important consideration is the intended use of the stairway. How many people will be using it at one time? You must know this answer to establish the width of the steps. The stairway shown above is 30 in. wide, just wide enough for its solo user to make it to the upper yard from his walkout basement. A garden stair for an arm-in-arm couple should be at least 48 in. wide.

Who will use the stairway? The answer will effect the riser height and the tread depth. Younger folks find it easy to ascend an 8-in. riser; older people find a 7-in. riser more comfortable; I usually build my steps with 7-in. risers. The stairway path will affect these dimensions, too. For example, if the steps pass through a garden, the cadence ought to be slower than if the steps serve only as access from one level to another. Normally, I make the tread 10 in. to 13 in. deep.

The height of the stairway is measured from one grade to the other. If the finished grade has

Interlocking details. **A dry-laid stairway requires careful fitting for both strength and beauty. The curved stairs and the stairwell walls are keyed together at every tread stone for additional strength. Photo by Robert Marsala.**

stairway, where I arrange for further drainage even in well-drained soils. In this project, I buried a 4-in. perforated pipe toward the outside of the stairwell foundation; it ties into the house's perimeter storm drain.

A permeable foundation—After placing a few scrap stones at the very bottom of the excavation, I dump in a 2½ ft. to 3-ft. thick layer of mixed ½-in. and ¾-in. crushed rock. Over this layer I add scrap stones of various sizes. Here's a tip: When your stone is delivered to the site, place it in a spot that won't hinder other construction and avoid underground soft spots, such as septic fields, buried gas tanks and electrical and drainage conduits.

The scrap stones make a sturdy base, which I further strengthen by working smaller chunks of stone into voids, then filling any remaining nooks and crannies with both ½-in. and ¾-in. crushed stone. Carefully working each chunk into a void strengthens the entire mass. More crushed stone is used as a filler only after the larger stones are laid.

All this stone packing creates a very strong and fairly solid mass. Water will drain completely through it, but silt and dirt won't penetrate too deeply. Silt and dirt hold moisture, and when that moisture freezes there's trouble for the stair. I came to this method of building dry-laid stairs several years ago after years of wrestling with landscaping cloth and screening (both materials eventually rot) and plastic, which is impermeable. Unlike these synthetic materials, stone lasts forever.

Step by step—Before setting exposed finished stones, brush away excess crushed rock to make a clean seat. The first finish stone I set is the bottom landing stone, which is level across its width but tipped very slightly from back to front to allow water to run away from the stair. I gauge the degree of tip by eye, but if you want to be precise, shoot for a slope of ¼ in. per ft. After setting the landing stone, I scratch lines on it with my chisel to indicate the face of the adjacent riser and the stairwell walls. I mark subsequent treads the same way.

I alternate risers and treads as I work my way up the stair. A typical tread stone chosen for the stair pictured here measured 2½-in. thick. That meant that the total height of the riser stones beneath it would be 4½ in., resulting in a rise of 7 in. It is important to set tread and riser stones so that individual riser heights do not vary more than ¼-in. Even a small break in cadence can cause someone to fall.

After choosing a couple of riser stones, I place one along the scratched line on the tread below, and then I shim it into place. Next, I fill the spaces behind the riser stone with scrap stone to stabilize it (photo following page). Although I build shims into the riser base, a few shims also may be required after the riser stones are in place. Placing just enough scrap stone into the spaces behind and underneath each stone to fill them keeps the riser stones from wobbling. Each successive riser stone must be firmed up in the same manner to make it

not quite been determined, as in this project, I can determine my riser and tread dimensions and tailor the grade to meet them. Where the grade is fixed, I'll divide the exact vertical dimension between grades by the number of risers to get the exact riser dimension.

The foundation—During the 22 years I've been working stone, I've tried several methods for building a foundation for dry-laid stone steps. My preferred method results in a base that is deep enough to prevent Jack Frost from upsetting

the steps while being permeable enough to drain water completely through and away from the steps. Here in western Massachusetts the frost line is 4 ft. deep, so my subcontractor excavates 4 ft. down from the finished grade at the bottom of the stair and slopes the subgrade up to about 2½ ft. below grade at the top of the stair (drawing next page). Because the subgrade has a relatively steep slope, water won't pool behind the top of the stair; that's why the excavation can be relatively shallow at this point. The sloped subgrade drains water to the bottom of the

Tidy infill. **This view of the base for the next tread shows the author's nook-and-cranny method of infilling with stones of various sizes. Belair begins by placing scrap rock, then fills the gaps with crushed stone to assure adequate drainage. Photo by the author.**

strong enough to hold the weight of the stone above it.

Treads and the stairwell—After each set of riser stones is in place and secured with shims, it's time to set a tread. I like to tie treads into the risers and the stairwell structure, so I look for a long, wide stone. But it isn't always possible to find one piece for the tread. If any landing or tread requires more than one stone, each one must butt tightly to its neighbor. In any case, I carefully move each tread piece into place, trying not to disturb the riser stones beneath. After setting the tread (whether one stone or more), I shim it just as I worked each riser stone into place. This time, though, it should tilt slightly toward the front of the stair to assure drainage. Careful shimming takes care of that.

Then I continue building the fill on the sides and rear of the step. As the stairwell walls go up, I build up the riser height of the next step. At the top of the stair, the fill is tapered to allow for the capstone.

I continue cutting, placing and infilling until the steps and most of the adjoining stairwell are completed. Then I set the adjoining wall. After completing the wall and the stairwell, I select capstones, usually larger stones, to finish off the top of each. By the time the stairwell reaches grade, it's only about a foot thick. A little work with shovel and rake fine-tunes the grade, and the project is ready for topsoil and plantings. □

Dick Belair is a stonemason in Charlemont, Mass.

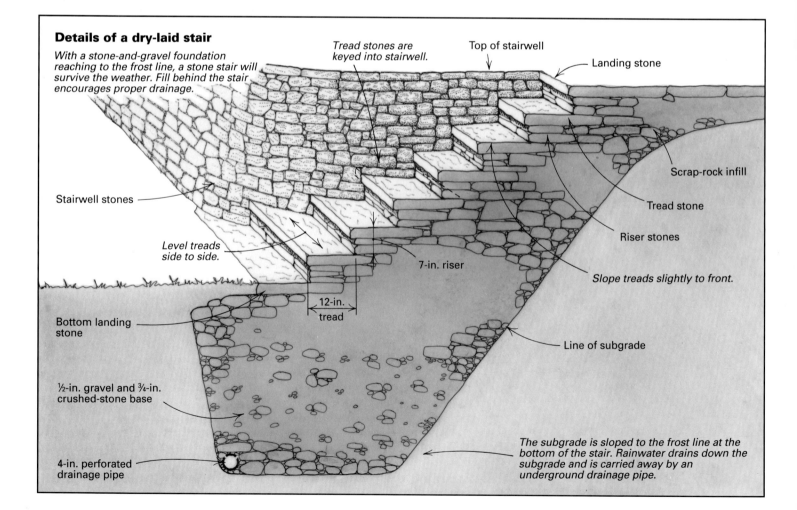

Details of a dry-laid stair

With a stone-and-gravel foundation reaching to the frost line, a stone stair will survive the weather. Fill behind the stair encourages proper drainage.

Tread stones are keyed into stairwell.

Top of stairwell

Landing stone

Stairwell stones

Level treads side to side.

Scrap-rock infill

Tread stone

Riser stones

7-in. riser

Slope treads slightly to front.

12-in. tread

Bottom landing stone

Line of subgrade

½-in. gravel and ¾-in. crushed-stone base

4-in. perforated drainage pipe

The subgrade is sloped to the frost line at the bottom of the stair. Rainwater drains down the subgrade and is carried away by an underground drainage pipe.

Drawing: Vince Babak

Photo: Karen Bussolini

Photo: Walter Smalling, Jr.

Plying the Trade

"Plywood," says stairbuilder Tom Luckey of Branford, Conn., "is my favorite building material." No doubt. Plywood veneered with cherry forms much of the stair at left. Typical of Luckey's stairs, it needed about five and one-half months of shop construction but only two weeks of site assembly (*Photo left by Ross Chapple*). The treads of the stair in the photo below are African mahogany plywood. They cantilever from a plywood box beam that forms the outside curve of the stair. Treads are supported by steel brackets welded to steel angles within the beam. Luckey's most abstract stair (photos right) is made from Zruynzeel, a solid marine-grade plywood imported from Holland. The stair above features a laminated-oak balustrade, built by Luckey, that spans the entire length of the stair.

Photo: Walter Smalling, Jr.

Photo: Karen Bussolini

Bill Reed,
W. G. Reed
Architecture,
Bethesda, Md.

Newel Posts

Sam and Marion Butz,
The Butz-Wilbern Partnership,
McLean, Va. *Photo: Kathy Buckalew*

Tony Farah, Big Twig Woodworks, Roosevelt, N. Y.

Thomas Frost, Jr.,
Frost Architecture,
Saratoga Springs, N. Y.

Brenda Burlin,
Sharpsburg, Md.

INDEX

The articles in this book originally appeared in *Fine Homebuilding* magazine. The date of first publication, issue number and page numbers for each article are given at right.